Analytical Measurement and Information

Advances in the information theoretic approach to chemical analyses

CHEMOMETRICS SERIES

Series Editor: **Dr. D. Bawden**
Pfizer Central Research, Sandwich, Kent, England

1. Distance Geometry and Conformational Calculations
G. M. Crippen

2. Clustering of Large Data Sets
Jure Zupan

3. Multivariate Data Analysis in Industrial Practice
Paul J. Lewi

4. Correlation Analysis of Organic Reactivity: With particular reference to multiple regression
John Shorter

5. Information Theoretic Indices for Characterization of Chemical Structures
Danail Bonchev

6. Logical and Combinatorial Algorithms for Drug Design
V. E. Golender *and* **A. B. Rozenblit**

7. Minimum Steric Difference
The MTD method for QSAR studies
Z. Simon, A. Chiriac, S. Holban, D. Ciubotaru *and* **G. I. Mihalas**

8. Analytical Measurement and Information
Advances in the information theoretic approach to chemical analyses
K. Eckschlager *and* **V. Štěpánek**

Analytical Measurement and Information

Advances in the information theoretic approach to chemical analyses

K. Eckschlager
and
V. Štěpánek
Czechoslovak Academy of Sciences, Czechoslovakia

RESEARCH STUDIES PRESS LTD.
Letchworth, Hertfordshire, England
JOHN WILEY & SONS INC.
New York · Chichester · Toronto · Brisbane · Singapore

RESEARCH STUDIES PRESS LTD.
58B Station Road, Letchworth, Herts. SG6 3BE, England

Marketing and Distribution:

Australia, New Zealand, South-east Asia:
Jacaranda-Wiley Ltd., Jacaranda Press
JOHN WILEY & SONS INC.
GPO Box 859, Brisbane, Queensland 4001, Australia

Canada:
JOHN WILEY & SONS CANADA LIMITED
22 Worcester Road, Rexdale, Ontario, Canada

Europe, Africa:
JOHN WILEY & SONS LIMITED
Baffins Lane, Chichester, West Sussex, England

North and South America and the rest of the world:
JOHN WILEY & SONS INC.
605 Third Avenue, New York, NY 10158, USA

Library of Congress Cataloging in Publication Data:

Eckschlager, Karel.
Analytical measurement and information.
(Chemometrics series; 8)
Bibliography: p.
Includes index.
1. Chemistry, Analytic—Mathematics. 2. Information theory. I. Štěpánek, Vladimír. II. Title. III. Series.
QD75.3.E35 1985 543 84-22331
ISBN 0 471 90652 2 (Wiley)

British Library Cataloguing in Publication Data:

Eckschlager, K.
Analytical measurement and information: advances in the information theoretic approach to chemical analyses.—(Chemometrics series; 8)
1. Chemistry, Analytic—Mathematics 2. Information theory
I. Title II. Štěpánek, V. III. Series
543'.001 QD75.3
ISBN 0 86380 021 1

ISBN 0 86380 021 1 (Research Studies Press Ltd.)
ISBN 0 471 90652 2 (John Wiley & Sons Inc.)

Printed in Great Britain

To the memory of our fathers

Editorial Preface

Since the introduction of the mathematical formalisations of information theory some decades ago, they have found application in many diverse fields of science and technology. They have proved to be particularly powerful when used to give quantitative expression for the information provided by a technique or apparatus of measurement. This monograph deals with such an application, in the field of analytical chemistry. The authors have been at the forefront of research in this area, and have already written a standard text, of which this monograph may be considered a continuation. The application of information theory to analytical chemistry is a topic likely to increase in importance in the future, and it is to be hoped that this monograph will inspire interest by analytical chemists in general, as well as those already involved in work in this area.

David Bawden
Sandwich, July 1984.

Preface

The aim of this monograph is to comprehensively summarize recent results achieved in the development of the information theory approach to the evaluation of methods, procedures and results in chemical analysis. Since we have published a fundamental treatise on this subject in our monograph 1979 (Information theory as applied to chemical analysis, Wiley-Interscience, New York) this new publication collects and discusses almost exclusively those results that have been obtained by us as well as by other scientists after that date. Knowledge of basic concepts of probability theory and of information theory are prerequisite for the readers or may be drawn from the quoted monograph, to which also the text of this publication refers in many places. Nevertheless a considerable extention of the theoretical apparatus appears in this book as a tool for treating new analytical problems.

The introductory chapter presents a survey of the development of information theory applied to analytical chemistry. It is followed by Chapter 1 in which the principles of obtaining analytical information are outlined. In Chapter 2 some information measures are briefly recalled, other ones are newly introduced and they are all discussed and compared. A broader theoretical framework has been transferred to the Appendix. Next,

Chapter 3 brings together results concerning information properties and values of information quantities obtained in various analytical procedures. It is divided into sections reflecting different types of chemical analyses and requiring different approaches to the evaluation. Chapter 4 deals with models of analytical systems and their impact upon information theoretic evaluation. Here sources of uncertainty are investigated and among them attention is paid to the calibration and to the effect of sampling techniques. A reader wishing to gain a deeper insight into the philosophy of measuring information and into the ties between various information measures adopted in this book is referred especially to Appendix A.4. References are given at the end of each chapter in alphabetical sequence.

We wish to thank the editor of the Chemometrics series Dr Bawden from Central Research, Pfizer Limited, Sandwich for his initiative with which he encouraged the writing of this book and for his reading the text in the preparatory period. Our thanks belong also to our colleagues A.Petřina, K.Baše, and J.Fusek for attention in typing the manuscript and assistance in conducting the illustrative material.

Prague, March 1984

K.Eckschlager
V.Štěpánek

Contents

Introduction

Although at the beginning of this century analytical chemistry was taken for a part of experimental chemistry and understood as a set of analytical methods and procedures based on chemical reactions, it has been developing, in recent decades, towards the utilization of various, mainly physical phenomena for analytical purposes. At the same time laboratory (instrumental and computing) technique continues its rapid advance. The analytical process accepts smaller quantities of material, it is possible to discriminate substances formerly undistinguishable and we can establish, by the means of local analysis, heterogenity of materials formerly treated as entirely homogeneous, etc. There occurs such conspicuous methodic differentiation of analytical chemistry that today individual methodologies develop altogether independently: they have their own "theoretical background", working methods, their own terminologies or at least laboratory slangs, and "methods" papers from different fields are published in closely specialized journals. Yet this differentiation, which greatly accelerated the development of analytical methods, narrows down the domain of interest of the analysts, sometimes to such an extent that they lose "the analytical approach" to problems to be solved.

Therefore the search for generally valid analytical points of view appears as a counterbalance to this differentiation: a major factor in these integration efforts has been the use of information theory in analytical chemistry.

Information theory has been developed since the end of World War II, first as a part of probability theory and today rather independently. Since information theory deals with the measurement of uncertainty, which is studied by the apparatus of the probability theory, this theory becomes fundamental for it. Information theory is pure mathematics, for it does not originate from empirical facts but from abstract definitions alone. Therefore, the statements of information theory apply everywhere in handling mathematical expressions of the same kind and its formulae do not depend implicitly on any empirical context.

The use of information theory in analytical chemistry has been subject, of course, to explicit development. In the first period at the beginning of the 1970s the members of "the Lindau circle" introduced general concepts and definitions from the point of view of system and information theories for the needs of analytical chemistry (Malissa, 1972). They also recognized that analytical chemistry has evolved into an independent scientific discipline dealing with methods obtaining and interpreting information about the chemical composition of material systems (Fresenius, 1977) or of signals bearing this information. Today information quantities are applied rather to evaluating, comparing, and optimizing analytical methods, procedures and devices (Eckschlager and Štěpánek, 1979, 1982; Liteanu and Rica, 1979; Cleij and Dijkstra, 1979; Danzer and Eckschlager, 1978; Danzer, 1973a, b, c, 1974, 1975a, b; Danzer and Marx, 1979; Eckschlager, 1975, 1976;

Frank et al., 1982; Danzer et al., 1982) and besides the "black-box" approach inherent in studies of information systems particular viewpoints are also of use which are conditioned by the chemical or physical substances of an analytical procedure or by properties of the device used. Also important is the application of information theory in decoding analytical signals, e.g., of IR or MS spectra (van Marlen and Dijkstra, 1976; Dupuis and Dijkstra, 1978). Kowalski (1980) classifies the use of information theory in analytical chemistry as a field of chemometrics.

Literature dealing with the use of information theory in analytical chemistry has become voluminous enough: one monograph (Eckschlager and Štěpánek, 1979) is entirely devoted to these problems and in another one (Doerffel and Eckschlager, 1981) we can also find application of information theory in the search for optimum strategy in analytical practice. Review articles (Cleij and Dijkstra, 1979; Eckschlager and Štěpánek, 1982; Frank et al., 1982; Liteanu and Rica, 1979) present surveys of some fields of analytical chemistry and of possibilities to adopt information theory in them. Another more than one hundred original papers can be divided into three groups:

1) Papers mostly originating from the beginning of the seventies, in which the importance of the use of information theory in analytical chemistry has been pointed out and in which some concepts and quantities have been defined by the use of the language and the relationships of information theory or of system theory. These papers have had great significance because they contribute to the modern and interdisciplinary concept of analytical chemistry as of an independent scientific discipline. From a variety of papers by the members of "the Lindau circle" the Malissas' textbook (1972) is to be commemorated preferably.

2) Papers that introduce information measures and quantities into analytical chemistry and draw attention to their use in analytical practice for the evaluation and optimization of analytical procedures. Originally mainly the Brillouin measure (Eckschlager, 1971; Danzer, 1973a, b, c; Danzer and Eckschlager, 1978; Danzer and Marx, 1979, 1982), more recently measures transferred from Shannon's communication theory have been introduced (Liteanu and Rica, 1979). Cleij and Dijkstra (1979) recommend equivocation and Frank et al. (1982) introduce transinformation for the evaluation of analytical results and methods. The divergence measure, which is a comprehensive measure of information obtained from measurements (Vajda and Eckschlager, 1980) was first adopted in chemical analysis by Eckschlager and Vajda (1974) and later its use was extended and generalized by Eckschlager (1975), Eckschlager and Štěpánek (1979, 1980, 1982).
3) Case studies, in which methods and relationships of information theory are used to evaluate and compare analytical results, methods, procedures and ways of processing analytical data, in choosing and optimizing analytical methods and procedures and in coding and decoding analytical signals (e.g., the spectra): over one hundred such papers have now been published.

Here we will try to summarize the main results of papers belonging to the second group, to classify the somewhat divergent definitions and terminology and to review the mathematical foundations of this domain, which forms today a field of chemometrics. Bonchev (1983) has reviewed the development and the application of information theory in various branches of chemistry in the introduction of his book published in Volume 5 of this Chemometrics Series. We explored some possibilities of the use of information theory in chemical analysis in our monograph of 1979. However, there exists a difference between chemometrics in analytical chemistry and in other branches of chemistry. Whereas in physical, organic, and inorganic chemistry or in biochemistry we mainly adopt prepared data that are reduced, classified or mutually related, the use of mathematical and statistical methods in analytical chemistry concerns almost exclusively with obtaining data, and therefore with a process of obtaining information about the chemical composition.

We have dealt in our monograph from 1979 as well as in this book only with information measures and possibilities of their practical employment in carrying out analyses or in theoretical analytical chemistry. Nevertheless we think that the aim of adopting chemometrical methods in analytical chemistry should be a simultaneous use of several mathematical methods that would supplement each other and that could be combined. If a specific problem is solved by the use of a number of mathematical methods at the same time, which have different substances and stem from different fundamental ideas we can reach, in favourable cases, a cooperation phenomenon, i.e., a synergetic tendency of individual methods to the given objective. This point of view has not been so far sufficiently pointed out in chemometrical literature, although a possibility to combine statistical methods with those of information theory in analytical practice in the search for optimum strategy has been mentioned in some examples in the monograph by Doerffel and Eckschlager (1981).

REFERENCES

Bonchev, D. (1983). Information Theoretic Indices for Characterization of Chemical Structures.

Research Studies Press, Chichester.
Cleij, P., and Dijkstra, A. (1979). Information theory applied to qualitative analysis. Fresenius Z. Anal.Chem. 298, 97-109.
Danzer, K. (1973a). Zu einigen Informationstheoretischen Aspekten der Analytik. Z.Chem. 13, 20-21.
Danzer, K. (1973b). Die Informationsmenge als Kenngrösse zweidimensionaler analytischen Aussagen. Z.Chem. 13, 69-70.
Danzer, K. (1973c). Ermittlung der Informationsmenge qualitativer Analysengänge. Z.Chem. 13, 229-231.
Danzer, K. (1974). Informationstheoretische Characterisierung der Verteiligungsanalysen. Z.Chem. 14, 73-75.
Danzer, K. (1975a). Zur Ermittlung der Informationsmenge bei spektrochemischen Analysenverfahren. Z.Chem. 15, 158-159.
Danzer, K. (1975b). Kennzeichnung der leichstungsvermögen des rentablen Einsatzes von Analysenverfahren mit Hilfe der Informationstheorie. Z.Chem. 15, 326-327.
Danzer, K., and Eckschlager, K. (1978). Information efficiency of analytical methods. Talanta 25, 725-726.
Danzer, K., and Marx, G. (1979). Zu einigen Informationstheoretischen Grundlagen der Strukturanalytik. Chem.Analit. 24, 33-42; 43-50.
Danzer, K., Hopfe, V., and Marx, G. (1982). Möglichkeiten der Erhöhung der Informationsmenge spektroskopischer Analysenmethode mit Hilfe der Rechentechnik. Z.Chem. 22, 332-338.
Doerffel, K., and Eckschlager, K. (1982). Optimale Strategien in der Analytik. Verlag Harri Deutsch. Frankfurt/M.

Dupuis, P.F., and Dijkstra, A. (1978). An information theoretical evaluation of the applicability of the ASTM Infrared Data Base for retrieval purposes. Fresenius Z.Anal.Chem. 290, 357-368.

Eckschlager, K. (1971). The amount of information obtained by analysis. Coll.Czechoslov.Chem.Commun. 36, 3016-3022.

Eckschlager, K. (1975). Informationsgehalt analytischer Ergebnisse. Fresenius Z.Anal.Chem. 277, 1-8.

Eckschlager, K. (1976). Zur Ermittlung der Informationsmenge bei zweidimensionalen Analysenverfahren. Z.Chem. 16, 111-112.

Eckschlager, K., and Štěpánek, V. (1979). Information theory as applied to chemical analysis. J.Wiley-Interscience, New York.

Eckschlagr, K., and Štěpánek, V. (1980). Accuracy of analytical results. Coll.Czechoslov.Chem.Commun. 45, 2516-2523.

Eckschlager, K., and Štěpánek, V. (1982). Information theory in analytical chemistry. Anal.Chem. 54, 1115A-1127A.

Eckschlager, K., and Vajda, I. (1974). Amount of Information of repeated higher precision analyses. Coll.Czechoslov.Chem.Commun. 39, 3076-3081.

Frank, J., Veress, G., and Pungor, E. (1982). Some problems of the application of information theory in analytical chemistry. Hungar.Sci.Instr. 54, 1-9.

Fresenius, W. (1977), in Reviews on Analytical Chemistry, Results and Trends in Analytical Chemistry during the Past 20 Years. p. 11. Akademiai Kiadó, Budapest.

Kowalski, B.R. (1980). Chemometrics. Anal.Chem. 52, 112R-122R.

Liteanu, C., and Rica, I. (1979). Utilization of the amount of information in evaluation of analytical

methods. Anal.Chem. 51, 1986-1995.
Malissa, H. (1972). Automation in und mit der Analytischen Chemie. Verlag der Wiener Medizinischen Akademie, Wien.
Vajda, I., and Eckschlager, K. (1980). Analysis of a measurement information. Kybernetika 16, 120-144.
van Marlen, G., and Dijkstra, A. (1976). Information theory applied to selection of peaks for retrieval of mass spectra. Anal.Chem. 48, 595-598.

CHAPTER 1
Obtaining Analytical Information

Analysis, as a process of obtaining information about the chemical composition of an analysed material system, takes place in a real probabilistic system with an input, an output, and an input-output relation. This system consists of two basic subsystems: in the first one (e.g., in a device) information arises and leaves it coded into an analytical signal and in the second one (e.g., in a computer) it is decoded. This decoding of a signal is an operation fundamental to the entire process. Sometimes it is merely the way of processing the signal which discriminates an instrumental analysis from a physico-chemical measurement, for instance from assessment of some material constants.

In a sample entering an analytical system qualitatively different components $\underline{i}$ (i=1,2, ... k) are present; their contents (concentrations) X_i in the sample are fixed but unknown quantities.

An analytical signal which leaves the first subsystem is either one-dimensional, i.e., it has only intensity η_i which, as a random variable, takes on continuous values y_{ij}(j=1,2, ... n) where $\underline{n}$ is the number of repeated measurements, or it is two-dimensional and the intensities of the signal are different in different positions. The position in which the signal assumes a conspicuous value for a particular component i (e.g., the maximum value of the peak or the half height of the wave) is a random variable taking on different values z_{ij} if the measurements are repeated. Its mean value $\bar{z}_i$ is either a physical constant (e.g., the wave length of the spectral line) or it depends on the course of the analysis (e.g., elution quantities in gas or fluid chromatography). An analytical method with one-dimensional signals makes possible a more or less specific identification, proof or determination of one component. A method with two-dimensional signals provides essentially the same selectively for several components simultaneously. In this case the position of the signal contains information about the identity of the analysed material (i.e., information about the qualitative composition) and its intensity in a given position supplies information about the quantitative content.

The result of an analysis is always a random variable; in identification analysis we determine the most likely identity and in quantitative analysis we seek for the mean value of the determinations for a particular analyte, or for several components at the same time.

The decoding of a signal in identification or qualitative analysis is carried out in such a way that positions $\bar{z}_i$ or ranks of the signals are assigned to individual identities i. Formally we can express this assignment in terms of an operator relation

$$i = A\left\{y_i \geqq y_{min} | \bar{z}_i\right\} \qquad (1.1)$$

where the operator A transfers the set of signals of intensity greater than y_{min} in the position $\bar{z}_i$ to the set of possible components of the analysed sample.

In decoding information about quantitative composition, we use the dependence of the average signal intensity on the content of the component to be determined, i.e., the analytical function

$$E[\xi_a] = f_A(E[\eta_a]) \quad (1.2)$$

Of course, we do not know this dependence, yet we can determine experimentally the calibration function

$$E[\eta_a] = f_K(X_a) \quad (1.3)$$

where X_a is the true (known) content of the analyte in the sample, in this case in the standard sample used for the calibration. The calibration function is inverse to the analytical function. We call the derivative

$$S_a = \frac{dE[\eta_a]}{dX_a} = \frac{df_K(X_a)}{dX_a} \quad (1.4)$$

the sensitivity. The formula (1.3) represents considerable simplification of the real dependence of the average signal intensity on the content of the analyte X_a and on the contents of remaining components $X_i(i=1,2, \ldots k;\ i \neq a)$

$$E[\eta_a] = f_K(X_a;\ X_1,\ X_2,\ \ldots,\ X_k;\ T_j) \quad (1.5)$$

This general dependence can be often written as

$$E[\eta_a] = f_{K,1}(X_a) + f_{K,2}(X_1,\ X_2,\ \ldots,\ X_k) + f_{K,3}(T_j) \quad (1.6)$$

where the term $f_{K,2}(X_1,\ X_2,\ \ldots,\ X_k)$ is denoted as the matrix-effect and the term $f_{K,3}(T_j)$ expresses the dependence on parameters T_j that characterize conditions under which the analysis is carried out. Obviously

$$\frac{\partial f_{K,2}(x_1, x_2, \ldots, x_k)}{\partial x_i} \ll s_a \quad i=1,2, \ldots, k;\ i \neq a$$

must hold; otherwise it is necessary to amend the matrix-effect or to choose another way of calibration.

In judging results of qualitative or identification analysis the selectivity is of importance, whereas we assess results of quantitative analysis according to their bias (the mean error $\delta = |X_i - E[\xi_i]|$) and to their precision (the variance $V[\xi_i]$). The precision and the unbiasedness of the results have an effect on the size of both the random and the systematic errors of the results. In practice the calibration conditions the attainment of the precision and of the unbiasedness of the determinations.

Further details on the analytical information process have been presented in our monograph (Eckschlager and Štěpánek, 1979) in Chapter 2.

REFERENCES

Eckschlager, K., and Štěpánek, V. (1979). Information theory as applied to chemical analysis. J.Wiley-Interscience, New York.

CHAPTER 2
Information Measures in Chemical Analysis

In the information theoretic approach to an analytical procedure, we usually wish to find the amount of information obtained by performing an analysis. This can be done by employing such different measures as those introduced by Brillouin, Shannon or Kullback; yet measures so far adopted for the needs of analytical chemistry represent, in principle, enough distinct tools for the given task.

In the sense of the Wiener definition, according to which we obtain information by decreasing the uncertainty, the amount of information need not to be linked with its semantic content nor with its truth. Yet it is useful if we know to what extent information obtained by an analysis is true. Individual measures of information content characterize more or less also the truth, i.e., the coincidence of an obtained result with the real chemical composition of the analysed sample or of the material from which the sample was taken. In the case of identification analyses we want the identity i_A obtained from an analysis to coincide with the real identity of the component present in the sample, i.e., that the probability inequality

$$1 - \varepsilon \leqq P(i_A = i) \leqq 1, \qquad \varepsilon \ll 1$$

hold and, in the case of quantitative analyses, we

require the value of the content or of the concentration μ_A found by the analysis to coincide with the true content X, i.e., that the probability

$$P(|\mu_A - X| < \varepsilon) \approx \alpha$$

for a value ε determined, for instance, as a statistically insignificant critical value of the difference on a level α.

Thus in judging individual measures of information content we can also take into account to what extent they satisfy these requirements concerning the size of information content I expressed by them:

1) Identification analysis:
   (i) $I \geqq 0$ for $1 - \varepsilon \leqq P(i_A = i) \leqq 1$, $\varepsilon \ll 1$;
   (ii) In comparing two different analytical methods the inequality $I_1 > I_2$ holds for $P_1(i_A = i) > P_2(i_A = i)$.
2) Quantitative analysis:
   (i) $I \geqq 0$ for $P(|\mu_A - X| < \varepsilon) \approx \alpha$;
   (ii) $I_1 > I_2$ for $\varepsilon_1 < \varepsilon_2$ where both values of ε are determined for the same level α.

In next sections we will treat the mentioned three information measures separately.

2.1 BRILLOUIN MEASURE

In 1928 Hartley expressed the information content of an element of a set E_N having N elements by $I(E_N) = \text{ld}\, N$ where ld is the binary logarithm (logarithm to base 2). In deriving this measure he started from following three postulates:

1. $I(E_N) \leqq I(E_{N+1})$
2. $I(E_{NM}) = I(E_N) + I(E_M)$ for $N, M = 1, 2, \ldots$ (2.1.1)
3. $I(E_2) = 1$

Postulate 2 (of the additivity) is understood in such a way that a set E_{NM} can be decomposed into N subsets each of M elements. This information measure, which was taken over and extended by Brillouin (1964), is suitable for use in analytical chemistry because of its simplicity in the case of discrete, qualitatively distinct, identities and it stands as the basis of Kaiser's quantity "informing power of an analytical method". The definition of information according to Brillouin is only defined in the case of n_o equally likely identities before analysis and $\underline{n}$ equally probable identities after analysis; it is given by $I = - \mathrm{ld}\,(n/n_o)$.

For continuous probability distributions, Brillouin approximates the number of possible discriminations as a ratio of the widths of confidence intervals of the a priori and the a posteriori probability distributions. This approach has been used, for instance, by Doerffel and Hildebrandt (1969) who evaluated the information content of the results of a quantitative analysis as

$$I = \mathrm{ld}\,\frac{(x_2 - x_1)\sqrt{n_p}}{2s_x t(\alpha, f)} \qquad (2.1.2)$$

where x_1, x_2 are the lower and the upper limits respectively within which the content of the analyte is anticipated, s_x is the sample standard deviation of the determinations and n_p is the number of parallel analyses. Doerffel chooses the critical value of the Student distribution $t(\alpha, f)$ with $f=n-1$ on the significance level $\alpha=0.05$. The Brillouin measure was adopted also by Eckschlager in an early paper (1971), before passing to the use of the divergence measure having deeper mathematical background and broader domain of applications. The amount of information obtained from results subject to a systematic error δ can be treated by the use of non-central t-distribution as

$$I = \mathrm{ld}\,\frac{(x_2 - x_1)\sqrt{n_p}}{2s_x t(\alpha, f, \delta)} \qquad (2.1.3)$$

Recently Danzer et al. (1982) have employed the Brillouin measure in discussing the possibilities of enhancing the amount of information obtained by a computer analysis of a two-dimensional signal in spectral analysis. They have derived

$$I = N_z\ \mathrm{ld}\ N_y \qquad (2.1.4)$$

with

$$N_z = \frac{z_{max} - z_{min}}{\Delta z} \qquad (2.1.5a)$$

and

$$N_y = \frac{(y_{max} - y_{min})\sqrt{n_p}}{2s_y t(\alpha, f)} \qquad (2.1.5b)$$

being the numbers of distinguishable signal positions and intensities respectively.

Compared with statistical evaluation of analytical results the Brillouin measure

1. takes into account the a priori and the a posteriori uncertainties expressed either as numbers of discrete identities or as widths of intervals; the ratio of these uncertainties represents the number of

discriminations.

2. introduces the logarithmic scale. The base of the logarithms determines the information units (e.g., bits).

2.2 SHANNON'S MEASURE

Next we need to measure the amount of information provided by the result of a random experiment that has n possible outcomes with probabilities $p_i = P(X_i)$ $(i=1,2, \ldots, n)$. It is in fact an extension of the previous case when the set (the sample space) is partitioned into n subsets: the elements then belong to them with probabilities p_i; we examine the amount of information necessary to know to which subset an element belongs. The sole measure that satisfies a few reasonably formulated assumptions (see for details, e.g., Chapter 4 of our monograph (1979)) is Shannon's entropy

$$I = - \sum_{i=1}^{n} p_i \operatorname{ld} p_i \qquad (2.2.1)$$

which was introduced in connection with the development of communication theory (Shannon, 1948, 1949). In the case where $p_i=1/n$ for all i $(i=1,2, \ldots, n)$ this measure reduces to Hartley's formula.

A quantity computed according to (2.2.1) can be interpreted also as uncertainty $H(P(X))$ before an experiment (e.g., before an identification analysis complying with the introduced model). We put $-0 . \operatorname{ld} 0 = 0$ for $p_i=0$. Moreover

$\sum_{i=1}^{n} p_i = 1$ and $0 \leqq H(P(X)) \leqq \operatorname{ld} n$, in which the

uncertainty is zero if one $p_i=1$ and the others equal zero (a sure event) and it equals the maximum value ld n when all probabilities are equal to 1/n.

Shannon's measure can be under particular assumptions extended to infinitely many discrete outcomes of a random experiment and it is defined for a continuous random variable with a probability density $p(x) > 0$ as

$$H(p(x)) = - \int_{-\infty}^{\infty} p(x) \,\mathrm{ld}\, p(x) dx \qquad (2.2.2)$$

where $\int_{-\infty}^{\infty} p(x) dx = 1.$

The communication theory built in the forties is based on the notion of two random variables ξ and η, which are generally stochastically dependent, and on seeking for what information about one random variable can be provided by the other. The analogy between the communication system formulated in this way and an analytical system is obvious: ξ can be the identity or content or concentration of an analyte and η can be the position or intensity of a signal. Thus in analytical chemistry it is possible to adopt the manner of evaluation introduced by Shannon for communication systems. It is accompanied by the following problems:

1. How can an analytical system be characterized as a communication system?
2. What models can be postulated for the description of a communication system?
3. What probability distributions of signals can be supposed in describing communication systems?
4. What quantity is to be used for measuring uncertainty arising in coding, transfer, and decoding analytical information?

An analytical system consists of two subsystems: in the first subsystem the analytical information arises and is coded into a signal while in the second one the signal is decoded into information about the composition. Coding can be understood as assignment of the signal position to an identity

$$E[\xi_i] = A(i)$$

or of the signal intensity to the given concentration

$$E[\eta_i] = B(x_i)$$

and the decoding as inverse assignment

$$i = A^{-1}(E[\xi_i])$$

or

$$x_i = B^{-1}(E[\eta_i])$$

We can take concentration or identity to be a random variable ξ, and position ζ and intensity η of the signal are generally also random variables; so that we always have at least two random variables, of which η or ζ supply information about ξ. We can then express the uncertainty after analysis in terms of conditional probabilities, for instance as compound entropy in identification or qualitative analysis (a discrete case)

$$H(P(X|Z_j)) = -\sum_{i=1}^{n} P(X_i|Z_j) \text{ ld } P(X_i|Z_j)$$

$$j = 1, 2, \ldots, m \qquad (2.2.3)$$

where $P(X_i|Z_j)$ is the conditional probability of the

presence of the component X_i after obtaining a signal in position Z_j in the output (Cleij and Dijkstra, 1979). The quality of the whole analytical procedure can be judged by the average uncertainty after analysis

$$H(P(X|Z)) = \sum_{j=1}^{m} P(Z_j)H(P(X|Z_j)) \qquad (2.2.4)$$

which we call equivocation.

Whereas the input-output relation in analyses of this kind furnishes conditional probabilities $P(Z_j|X_i)$ the probabilities $P(X_i|Z_j)$ appearing in (2.2.3) are obtained from the well known Bayes' theorem as

$$P(X_i|Z_j) = \frac{P(X_i).P(Z_j|X_i)}{P(Z_j)}$$

where $P(Z_j)$ is the probability of measuring a signal in position Z_j and can be computed by the formula of total probability

$$P(Z_j) = \sum_{i=1}^{n} P(X_i).P(Z_j|X_i)$$

If we consider the model of the analysis as a channel without noise, or with negligible noise, we can measure the amount of information as a decrease of uncertainty, i.e., the difference of uncertainties before and after analysis expressed as Shannon's entropies. Thus, e.g., in the case of the signal output in position Z_j

$$I = H(P(X)) - H(P(X|Z_j)) \qquad (2.2.5)$$

When we turn our attention to continuous cases we can assign a uniform probability distribution $p(x)=1/(x_2-x_1)$ to the unknown concentration within an interval $\langle x_1, x_2 \rangle$ ($p(x)=0$ outside the interval) - lacking any knowledge as pre-information - and the a posteriori entropy will employ the normal distribution $N(\mu, \sigma^2)$ of the results of the determinations. Then the amount of information computed as the decrease of uncertainty turns out to be

$$I = \mathrm{ld} \frac{x_2 - x_1}{\sigma\sqrt{2\pi e}} \tag{2.2.6}$$

(see, e.g., Eckschlager, 1975). If we compare this information measure with (2.1.2) it is obvious that for $n_p=1$ and $t(\alpha,\infty)=\sqrt{2\pi e}/2$ (which happens if $\alpha=0.0388$) both formulae yield the same value.

The disadvantage of the information measure computed as the decrease of uncertainty in (2.2.5) is that it can in some instances become negative. However, it can be shown that by substracting the equivocation (2.2.4) we always get non-negative information

$$I = H(P(X)) - H(P(X|Z)) \tag{2.2.7}$$

Information expressed in this way is a specific case of another information quantity called transinformation or mutual information, i.e., information about one random variable contained in the determination of another random variable (Aczél and Daróczy, 1975). Details on these connections are given in the Appendix at the back of this book.

The search for analogies between an analytical and a communication system, and the employment of Shannon's communication theory in analytical chemistry (Liteanu and Rica, 1979; Cleij and Dijkstra, 1979; Frank et al., 1982) are still of importance, above all in qualitative analysis. However, the difference of entropies can be negative and does not express to what extent the result of the experiment confirms our preliminary assumptions. Besides in the course of time several quantities originating from communication theory have been used in analytical chemistry which are sometimes not uniformly defined, denoted and interpreted and, futhermore, we do not find in any model of a communication system adequate analogies with such features of analytical systems as the bias or the matrix-effect. Thus the analogy between these systems was certainly expedient in the beginnings of introducing information theory for analytical usage; today we need not manage with such analogies any longer.

2.3 DIVERGENCE MEASURE

Analysis can be understood as the process of obtaining information, i.e., of reducing the original (a priori) uncertainty of our knowledge of the chemical composition (an identity or a fixed but unknown content of an analyte) and replacing it by a lesser a posteriori uncertainty of the knowledge of the same variable. Since, as we have seen, the difference of entropies can be negative whereas the amount of information cannot, it is desirable to find a non-negative information measure. Such a measure is the quantity $I(q||p)$ called by Rényi (1970) the information gain and derived for instance in Section 4.3 of our monograph (1979). In that publication we make extensive use of that measure in evaluating information contents obtained in various analytical procedures and in considering its application to different analytical methods. Only in the specific case when we transfer from a uniform distribution to any other probability distribution does this measure equal the difference of entropies (cf. page 79 of our monograph, 1979). Kullback has considered it as a definition of information for discriminating in favour of one hypothesis against another hypothesis (Kullback, 1959).

In a finite discrete case when we replace probabilities $p_i = p(x_i)$ by probabilities $q_i = q(x_i)$ the gain of information is

$$I(q||p) = \sum_{i=1}^{n} q_i \operatorname{ld} \frac{q_i}{p_i} \tag{2.3.1}$$

Under assumptions of convergence this measure can be extended to the infinite number of discrete values. The derivation for continuous distributions along with necessary assumptions will be found in our monograph. It results in

$$I(q||p) = \int_{-\infty}^{\infty} q(x) \operatorname{ld} \frac{q(x)}{p(x)} dx \tag{2.3.2}$$

where we pass from a probability density $p(x)$ to a density $q(x)$.

Another interpretation of the information gain than as a measure of the divergence of the a priori and the a posteriori distributions is possible. The inaccuracy of an assertion that a continuous random variable has a probability density $p(x)$ while the true distribution is $q(x)$ is characterized by the Kerridge-Bongard measure (Kerridge, 1961; Bongard, 1963)

$$H(p|q) = - \int_{x_1}^{x_2} q(x) \operatorname{ld} p(x) dx \tag{2.3.3}$$

where $p(x) > 0$ for $x \in \langle x_1, x_2 \rangle$ and

$$\int_{x_1}^{x_2} p(x)\,dx = \int_{x_1}^{x_2} q(x)\,dx \approx 1$$

This measure satisfies a set of assumptions similar to those used in the derivation of the entropy. We can find that

$$H(p|q) = H(q) + I(q||p)$$

Thus this measure is a sum of uncertainty linked with the distribution $q(x)$ and of the error in the anticipation of this distribution. From this point of view we can designate $I(q||p)$ as "the error term".

Therefore if $p(x)$ describes our knowledge of the content of an analyte prior to analysis, and $q(x)$ is the a posteriori probability density of the results of the analysis, the information gain can be evaluated as

$$I(q||p) = H(p|q) - H(q) = \int_{x_1}^{x_2} q(x)\ \mathrm{ld}\,\frac{q(x)}{p(x)}\,dx \qquad (2.3.4)$$

For arbitrary $p(x)$ and $q(x)$ we have $H(p|q) \geqq H(q)$; if $p \equiv q$, the sign of equality is valid.

This interpretation of the divergence measure in the case of its use for evaluating the information content of the results of a quantitative analysis has its own rationale: we face a situation such that, before performing analysis, we assume the content of the analyte in the sample to have a distribution $p(x)$ and that the experiment supplies its real distribution $q(x)$ afterwards. Such a priori inaccuracy is expressed by the Kerridge-

Bongard measure while the a posteriori uncertainty is given by Shannon's entropy.

We can easily show (Eckschlager and Štěpánek, 1982) that the divergence measure of the information content in (2.3.1) and (2.3.2) has following properties:

1. $I(q||p) \geqq 0$, i.e., it is non-negative and equal to zero if $p(x) \equiv q(x)$.
2. $I(q_1||p) > I(q_2||p)$ if the expectations of both a posteriori distributions are equal, i.e., $\mu(q_1)=\mu(q_2)$ but the standard deviations are in relation $\sigma(q_1) < \sigma(q_2)$; the pre-distribution $p(x)$ is the same in either case.
3. $I(q||p_1) > I(q||p_2)$ if $\mu(p_1)=\mu(p_2)$ and $\sigma(p_1) > \sigma(p_2)$; the a posteriori distribution $q(x)$ does not vary.
4. If $\sigma(p_1)=\sigma(p_2)$ and $\sigma(q_1)=\sigma(q_2)$ then $I(q_1||p_1) > I(q_2||p_2)$ provided that $|\mu(p_1)-\mu(q_1)| > |\mu(p_2)-\mu(q_2)|$. The difference between $I(q_1||p_1)$ and $I(q_2||p_2)$ can be very small if $p(x)$ is constant in the whole interval $\langle x_1, x_2 \rangle$.

The content of those properties is obvious. Discussion and comparisons can be found in the quoted paper.

As far as the distribution of the results is true except that it is not unbiased (the presence of a systematic error) and the true distribution is $r(x)$ whose $\mu=X$ and $\sigma_r=\sigma_q$, then the information gain is given by a difference of Kerridge-Bongard quantities

$$I(r;q,p) = H(p|r) - H(q|r) = \int_{x_1}^{x_2} r(x) \operatorname{ld} \frac{q(x)}{p(x)} dx \quad (2.3.5)$$

This measure can lead to a negative value if incorrect results supply us with misleading information, i.e., it satisfies the condition 2(i) because it takes on negative values only then unless the condition $P(|\mu_A - X| < \varepsilon) \approx \alpha$

is fulfilled (where ε is determined as the boundary difference between the true and the found values that is not statistically significant on the level $\mathcal{L}$). This measure satisfies also the condition 2(ii) when adopted to quantitative analyses. We will come back to it in Chapter 3.

REFERENCES

Aczél, J., and Daróczy, Z. (1975). On measures of information and their characterizations. Acad.Press, New York.

Bongard, M.M. (1966). On the concept of "useful information" (in Russian). Problemy Kibernetiki 6, 91-130.

Brillouin, L. (1964). Scientific uncertainty and information. Acad.Press, New York.

Cleij, P., and Dijkstra, A. (1979) Information theory applied to qualitative analysis. Fresenius Z.Anal.Chem. 298, 97-109.

Danzer, K., Hopfe, V., and Marx, G. (1982). Möglichkeiten der Erhöhung der Informationsmenge spektroskopischer Analysenmethode mit Hilfe der Rechentechnik. Z.Chem. 22, 332-338.

Doerffel, K., and Hildebrandt, W. (1969). Beurteilung von Analysenverfahren unter Einsatz der Informationstheorie. Wiss.Z.Tech.Hochschule Leuna-Merseburg 11, 30-35.

Eckschlager, K. (1971). The amount of information obtained by analysis. Coll.Czechoslov.Chem.Commun. 36, 3016-3022.

Eckschlager, K. (1975). Informationsgehalt analytischer Ergebnisse. Fresenius Z.Anal.Chem. 277, 1-8.

Eckschlager, K., and Štěpánek, V. (1979).
Information theory as applied to chemical analysis.
J.Wiley Interscience, New York.

Eckschlager, K., and Štěpánek, V. (1982).
Some properties of the divergence measure of information content as related to quantitative analyses. Coll.Czechoslov.Chem.Commun. 47, 1195-1202.

Frank, I., Veress, G., and Pungor, E. (1982).
Some problems of the application of information theory in analytical chemistry.
Hungar.Sci.Instr. 54, 1-9.

Kerridge, D.F. (1961). Inaccuracy and inference.
J.Roy.Statist.Soc. Ser.B. 23, 184-194.

Kullback, S. (1959). Information theory and statistics.
J.Wiley, New York.

Liteanu, C., and Rica, J. (1979).
Utilization of the amount of information in evaluation of analytical methods.
Anal.Chem. 51, 1986-1995.

Rényi, A. (1970). Probability theory.
Akad.Kiadó, Budapest.

Shannon, C.E. (1948).
The mathematical theory of communication.
Bell.Syst.Tech.J. 27, 379-423; 623-656.

Shannon, C.E., and Weaver, W. (1949).
The mathematical theory of communication.
University of Illinois Press, Urbana.

CHAPTER 3
Information Theory Approach to Various Analytical Procedures

3.1 QUALITATIVE AND IDENTIFICATION ANALYSIS

Qualitative and identification chemical analyses by their nature require information-theoretical evaluation while statistical methods - in contrast to the situation with quantitative analysis - are less useful. Several authors have dealt with information-theoretical approach to evaluating these methods, especially A.Dijkstra and his co-workers from the State University in Utrecht. Some ways of approaching these tasks will be shown next.

In the case of instrumental or chromatographic qualitative or identification analysis, the input into the analytical system is given by a set of discrete qualitatively distinct identities $\underline{i}$ ($i=1,2, \ldots ,n_0$) present in the analysed sample in concentrations X_i. In the output we have a set of signals in different positions known beforehand z_j ($j=1,2, \ldots, m;\ m \geqq n$) where $\underline{n}$ is the number of components that we can prove simultaneously. A qualitative proof is sometimes feasible also if we know only the rank of signals corresponding to individual components and not their precise positions z_j. The input-output relation, i.e., the practical realization of the operator A from the relation (1.1) is described by a matrix of conditional probabilities a_{ji}

of dimension m.n

$$||a_{ji}|| = \begin{Vmatrix} a_{11} & a_{12} & \cdots & a_{1n} \\ a_{21} & a_{22} & \cdots & a_{2n} \\ \vdots & & & \\ a_{m1} & a_{m2} & \cdots & a_{mn} \end{Vmatrix}$$

Elements of this matrix $a_{ji} = P(Z_j|X_i)$ are conditional probabilities of obtaining a signal in position Z_j provided that the component i in concentration X_i is present in the input. For the objectives of qualitative analysis it is sufficient to distinguish three domains of concentrations bounded by values $x_{o,i}$ (the highest concentration for which we never obtain a signal and when $P(Z_j|X_i) = 0$ for $X_i \leqq x_{o,i}$) and $x_{1,i}$ (the lowest concentration for which we always get a signal of intensity $y_{ij} \geqq y_{min}$ and when $P(Z_j|X_i) = 1$ for $X_i \geqq x_{1,i}$). Yet for analytical needs we will rather employ conditional probabilities $P(X_i|Z_j)$ of the presence of the component i in concentration $X_i > x_{o,i}$ in the input, provided that we have obtained a signal in position Z_j in the output. These probabilities will be determined by Bayes' rule (cf. Section 2.2) by means of probabilities $P(Z_j|X_i)$, which we determine experimentally as relative frequencies of the occurence of a signal in the presence of the sole component X_i for various concentration levels (Liteanu and Rica, 1979). For the Bayes' theorem

$$P(X_1|Z_j) = \frac{P(X_1)P(Z_j|X_1)}{\sum_{i=1}^{n} P(X_i)P(Z_j|X_i)}$$

where X_1 is the analyte we need to know probabilities $P(X_i)$ from pre-information and probabilities $P(Z_j|X_i)$.

If a signal also arises in position Z_j in the presence of other components i (i=2,3, ..., n) then $0 < P(X_1|Z_j) < 1$ and the closer it is to one the more selective is the proof. The matrix of these probabilities of dimension n.n is, for a perfectly selective proof of all n components out of the total number n_o of components that can be present in the input ($n \leqq n_o$), the unit diagonal, i.e., $P(X_i|Z_i) = 1$ and $P(X_i|Z_j) = 0$, $i \neq j$.

The uncertainty of a proof or of identification of components i (i=1,2, ..., n) when a signal arises in position Z_j can be expressed as compound entropy from (2.2.3). The amount of information obtained from a signal in position Z_j, evaluated as the decrease of uncertainty, follows by substituting in (2.2.5)

$$I = -\sum_{i=1}^{n} P(X_i)\,\mathrm{ld}P(X_i) + \sum_{i=1}^{n} P(X_i|Z_j)\,\mathrm{ld}P(X_i|Z_j) \qquad (3.1.1)$$

This amount of information is a maximum for $H(P(X|Z_j))=0$, i.e., for an unambiguous proof of a particular component and unambiguous exclusion of the presence of all other ones; then it takes the value

$$I = -\sum_{i=1}^{n} P(X_i)\ \mathrm{ld}\ P(X_i)$$

As we have seen in (2.2.4) the quality of the whole analytical procedure can be described by equivocation (Cleij and Dijkstra, 1979)

$$E = H(P(X|Z)) = -\sum_{j=1}^{m} P(Z_j)H(P(X|Z_j))$$

$$= -\sum_{i=1}^{n}\sum_{j=1}^{m} P(Z_j)P(X_i|Z_j) \text{ ld } P(X_i|Z_j) \qquad (3.1.2)$$

However, because $P(Z_j)P(X_i|Z_j) = P(X_i,Z_j)$, i.e., the joint probability, we can now write

$$E = -\sum_{i=1}^{n}\sum_{j=1}^{m} P(X_i,Z_j) \text{ ld } P(X_i|Z_j) \qquad (3.1.3)$$

Similarly we can evaluate a procedure in qualitative or identification analysis by the amount of information

$$I = H(X) - E$$

$$= -\sum_{i=1}^{n} P(X_i) \text{ ld } P(X_i) + \sum_{i=1}^{n}\sum_{j=1}^{m} P(X_i,Z_j) \text{ ld } P(X_i|Z_j)$$

If we substitute $P(X_i)$ according to the formula of total probability as

$\sum_{j=1}^{m} P(X_i|Z_j)$ we get

$$I = \sum_{i=1}^{n}\sum_{j=1}^{m} P(X_i,Z_j) \text{ ld } \frac{P(X_i|Z_j)}{P(X_i)}$$

$$= \sum_{i=1}^{n}\sum_{j=1}^{m} P(Z_j)P(X|Z_j) \text{ ld } \frac{P(X_i|Z_j)}{P(X_i)} \qquad (3.1.4)$$

Because this is the mean value of the non-negative divergence measure

$$\sum_{i=1}^{n} P(X_i|Z_j) \operatorname{ld} \frac{P(X_i|Z_j)}{P(X_i)}$$

the formula in (3.1.4) yields always non-negative values.

The importance and the way of computing the equivocation are shown in the following example.

Example 3.1.1:

In TLC qualitative chromatography of 4 components to be expected with equal probabilities we obtain Gaussian signals with a constant standard deviation $\sigma=0.02$ R_f values. The measurement step of R_f is q=0.05. The averages of distributions of R_f's for each component are:

X_i	$\overline{Z}_j$ (in R_f values)
X_1	0.20
X_2	0.25
X_3	0.30
X_4	0.35

Thus the selection is poor and a perfectly selective proof is impossible. We determine probabilities $P(Z_j|X_i) = \Phi(\overline{Z}_j + q/2) - \Phi(\overline{Z}_j - q/2)$ where Φ is the distribution function of the normal distribution $N(\overline{Z}_j, \sigma^2)$. In terms of the Bayes' rule we calculate probabilities $P(X_i|Z_j)$ and compile matrices of $P(Z_j|X_i)$ and $P(X_i|Z_j)$. For each row of the latter we compute

$$P(Z_j)H(P(X|Z_j)) = - P(Z_j) \sum_{i=1}^{n} P(X_i|Z_j) \operatorname{ld} P(X_i|Z_j)$$

and the sum of these terms is the equivocation E.

TABLE 1. Matrix of conditional probabilities $P(Z_j|X_i)$

Z_j	X_i			
	X_1	X_2	X_3	X_4
0.15	0.105	0	0	0
0.20	0.790	0.105	0	0
0.25	0.105	0.790	0.105	0
0.30	0	0.105	0.790	0.105
0.35	0	0	0.105	0.790
0.40	0	0	0	0.105

TABLE 2. Matrix of conditional probabilities $P(X_i|Z_j)$

Z_j	X_i				$P(Z_j)H(P(X|Z_j))$
	X_1	X_2	X_3	X_4	
0.15	1	0	0	0	0
0.20	0.883	0.117	0	0	0.116
0.25	0.105	0.790	0.105	0	0.238
0.30	0	0.105	0.790	0.105	0.238
0.35	0	0	0.117	0.883	0.116
0.40	0	0	0	1	0
					E = 0.708 bits

Because the uncertainty prior to the experiment was 2 bits it turns out that this analytical method reduces the uncertainty by more than one half; of course, individual components cannot be identified without error. The amount of information obtained equals 1.106 bits.

A similar example has been introduced by Liteanu and Rica (1979).

For $X_i \geqq x_{1,i}$ neither entropy nor equivocation and information content depend on the true concentration X_i. However in the domain $x_{o,i} < X_i < x_{1,i}$ the entropy is dependent on X_i and for a perfectly specific proof we obtain the following: in the interval $\langle x_{o,1}; x_{1,1} \rangle$ where X_1 is the analyte and when $P(X_i) = 1/n$ $(i=1,2, \ldots, n)$ the probability $P(Z_j|X_1)$ moves between 0 and 1 while the remaining $P(Z_j|X_i) = 0$ for $i \neq 1$. Moreover $0 < P(\overline{Z}_j|X_1) < 1$ and $P(\overline{Z}_j|X_i) = 1$, $(i \neq 1)$ if $P(\overline{Z}_j|X_i)$ $(i=1,2, \ldots, n)$ denotes the probability of missing a signal if a component X_i is present. Whereas the conditional probability function when a signal appears is simple ($P(X_1|Z_j) = 1$, $P(X_i|Z_j) = 0$ if $i \neq 1$) the conditional distribution under absence of a signal turns out (from the Bayes' theorem)

$$P(X_1|\overline{Z}_j) = \frac{P(X_1) \cdot P(\overline{Z}_j|X_1)}{\sum_{i=1}^{n} P(X_i) P(\overline{Z}_j|X_i)}$$

$$= \frac{\frac{1}{n} P(\overline{Z}_j|X_1)}{\frac{1}{n}[n-1+P(\overline{Z}_j|X_1)]} = \frac{P(\overline{Z}_j|X_1)}{n-1+P(\overline{Z}_j|X_1)}$$

$$P(X_i|\overline{Z}_j) = \frac{1}{n-1+P(\overline{Z}_j|X_1)} \qquad i \neq 1$$

Since $P(\overline{Z}_j|X_i) = 1-P(Z_j|X_i)$ $(i=1,2, ..., n)$ these probabilities convert into

$$P(X_1|\overline{Z}_j) = \frac{1-P(Z_j|X_1)}{n-P(Z_j|X_1)} \qquad (3.1.5)$$

$$P(X_i|\overline{Z}_j) = \frac{1}{n-P(Z_j|X_1)} \qquad i \neq 1$$

Then the uncertainties after analysis appear as (introducing $P(Z_j|X_i)$)

$$H(P(X|Z_j)) = 0$$

$$H(P(X|\overline{Z}_j)) = \text{ld } [n-P(Z_j|X_1)] - \frac{1-P(Z_j|X_1)}{n-P(Z_j|X_1)} \text{ ld } [1-P(Z_j|X_1)] \qquad (3.1.6)$$

Specifically for $P(\overline{Z}_j|X_1) = P(Z_j|X_1) = 1/2$, which is the case for concentration $X_1 = (x_{o,1} + x_{1,1})/2$ (Liteanu and Rica, 1979), we get

$$H(P(X|\overline{Z}_j)) = \text{ld } (n-\tfrac{1}{2}) + \frac{\text{ld } 2}{2n-1}$$

And finally the equivocation results in

$$E = P(Z_j).0 + P(\overline{Z}_j).H(P(X|\overline{Z}_j))$$

$$= \frac{n-P(Z_j|X_1)}{n} \text{ ld } [n-P(Z_j|X_1)] - \frac{1-P(Z_j|X_1)}{n} \text{ ld } [1-P(Z_j|X_1)] \qquad (3.1.7)$$

Specifically again for $P(Z_j|X_1) = 1/2$

$$E = \frac{2n-1}{2} \, ld \, (2n-1) - (n-1) \, ld \, 2$$

Thus both the entropy in (3.1.6) and the equivocation in (3.1.7) depend on the number of components and on the probability of getting a signal when the analyte is present. For n=1 (only one analyte is present) we obtain $H(P(X|\bar{Z}_j)) = E = 0$. It can be shown that with the probability $P(Z_j|X_1)$ varying from zero to one the equivocation decreases from ld n to the value $\frac{n-1}{n}$ ld (n-1) $(n \geqq 2)$.

For practical needs it is advantageous if the limit $x_{1,1}$ for the analyte is as low as possible and if, by contrast, $x_{1,i} \gg x_{1,1}$ for $i \neq 1$. A low value of the detection limit expressed, for instance, by the high value of $pD_1 = -\log x_{1,1}$ depends to a great extent on the sensitivity, defined in Chapter 1. If $y_{1,min}$ is the minimum perceivable signal intensity then

$$pD_1 = -\log x_{1,1} = -\log y_{1,min} + \log S_1$$

is the greater the greater is the sensitivity given by (1.4).

Formulae (2.2.3) and (3.1.6) are useful for judging procedures of qualitative and identification analysis: entropy by (2.2.3) depends principally on the selectivity of the procedure and entropy by (3.1.6) depends rather on its sensitivity. For a procedure serving the needs of identification analysis when sufficient amount of the sample is available so that we can work with concentration $X_i \gg X_{1,i}$, the perfect selectivity is fundamental. However in qualitative analysis, when the number of possible identities n_o is usually not great, - often enough we find out whether a particular single component is or is not present - we have sometimes to prove low concentrations. Then a high sensitivity of the procedure is necessary, so that we may not work, as far as possible, in the domain from $x_{o,i}$ through $x_{1,i}$. In that case the total uncertainty of the result is enhanced by that component of uncertainty that is expressed by entropy in (3.1.6).

It is possible, and sometimes sufficient to evaluate the information content of the result of a qualitative analysis by the divergence measure. In the most simple case, when we regard both the a priori and the a posteriori distributions as discrete uniform so that $p(x_i) = 1/n_o$ and $q(x_i) = 1/n$, the divergence measure yields (cf. our monograph, 1979, Section 6.2)

$$I(q||p) = \sum_{i=1}^{n} \frac{1}{n} \text{ld} \frac{\frac{1}{n}}{\frac{1}{n_o}} = \text{ld} \frac{n_o}{n} \qquad (3.1.8)$$

where $\underline{n}$ is the number of possible but so far unidentified components after the analysis was carried out. If various combinations of the components are possible, $\underline{n}$ and n_o have to be computed by the use of combinatorial techniques.

In (3.1.8) we assumed that all the components were equally likely (by adopting uniform probability distributions). Otherwise we have to evaluate the information content as

$$I(q||p) = \sum_{i=1}^{n} P(X_i|Z_j) \text{ ld} \frac{P(X_i|Z_j)}{P(X_i)} \quad j=1,2, \ldots, m \qquad (3.1.9)$$

Information theory can be relatively easily applied to one-channel procedures of qualitative analysis, but application to a combination of one-channel procedures (a multichannel procedure) is much more difficult.

Although information quantities such as uncertainty

after analysis, equivocation, information content, etc. can be equally expressed in qualitative and indentification analysis - that is why both cases are usually treated simultaneously (Eckschlager and Štěpánek, 1982; Cleij and Dijkstra, 1979) - an important difference is the relevance of information for qualitative and identification analysis respectively. Whereas an identification analysis consists in determining the identity of a component present in the analysed sample, for qualitative analysis it is relevant to find out whether the concentration of a component given beforehand is greater or less than the detection limit, i.e., whether it belongs to the interval $\langle 0, x_{o,A} \rangle$ or whether $x_A \geqq x_{o,A}$. We will later encounter a similar approach to quantitative trace analysis in Section 3.4.

3.2 QUANTITATIVE ONE-COMPONENT ANALYSIS

For the information theoretic evaluation of quantitative analysis, the Brillouin measure employing confidence intervals was first adopted (see Section 2.1); therefore at that time the view was justified that this evaluation was not advantageous compared to the use of statistical methods. Only the general model of both qualitative and quantitative chemical analyses, based on the analogy between analytical process and process of transmitting information by a communication system, and mainly the introduction of the divergence measure (Eckschlager and Štěpánek, 1979, 1982; Eckschlager and Vajda, 1974; Eckschlager, 1975), represent more important contributions to theoretical analytical chemistry or to the practice of evaluating and judging analytical procedures, methods, and devices and to the choice of optimum analytical strategy.

The input into an analytical system in one-component

quantitative analysis is given by a fixed but unknown content X_i of the component to be determined (the analyte) in the analysed sample and in the output we obtain a signal of intensity η_i, which is by its nature a continuous random variable. The result of the quantitative analysis is another continuous random variable ξ_i; the aim is to find its mean value $E[\xi_i]$, which is supposed to accord with X_i as much as possible. The input-output relation of an analytical system is given by the calibration and the analytical functions (see Chapter 1). The a priori distribution reflects preliminary assumptions or pre-information obtained, e.g., from preliminary semiquantitative determinations, and the a posteriori probability distribution is that of signal intensities or of results. The a posteriori distribution of the results is always closely linked with the distribution of signal intensity; in the most simple case of linear calibration, or an analytical function passing through the origin, the expected (mean) value of the result

$$E[\xi_i] = \frac{1}{S_i} E[\eta_i]$$

where S_i is the sensitivity by (1.4) and the variance

$$V[\xi_i] = \frac{1}{S_i^2} V[\eta_i]$$

Practical consequences of these relationships are discussed in Chapter 4 of (Doerffel and Eckschlager, 1981).

Only exceptionally is the relation between $E[\xi_i]$ and X_i, i.e., the calibration function, known beforehand, for instance from stoichiometry; in most cases it is determined experimentally by calibrating and its nature is that of regression relationship. The precision with which the calibration function is determined depends on the way in which the calibration is carried out, on the number of standards used, etc.; details can be found, e.g., in (Eckschlager and Štěpánek, 1979; Doerffel and Eckschlager, 1981).

Some comparisons of various information measures and quantities such as the difference of entropies, equivocation or transinformation have been introduced in Section 2.2 or are fully described in the Appendix. For evaluating results and methods of quantitative analysis the most important is the use of the information gain (the divergence measure) introduced in Section 2.2 and largely adopted in our monograph (1979). This quantity expresses "measurement information" in that way that it employs the statistical distribution generated by a random sample as the a posteriori distribution.

For the most frequent case of the a priori uniform distribution $U\langle x_1, x_2\rangle$ and the a posteriori normal distribution $N(\mu, \sigma^2)$ the information content expressed by a divergence measure is given for $x_1 + 3\sigma \leq \mu \leq x_2 - 3\sigma$ by the value in (2.2.6)

$$I(q||p) = \frac{x_2 - x_1}{\sigma\sqrt{2\pi e}}$$

If the results are means of $n_A \geqq 2$ parallel determinations we substitute $\sigma = \sigma_x/\sqrt{n_A}$. When the parameter σ_x is not known and is estimated by

$$s_x = \left[\frac{1}{n_s - 1}\sum_{i=1}^{n}(x_i - \bar{x})^2\right]^{1/2}$$

we compute (see Eckschlager and Štěpánek, 1979)

$$I(q||p) = \ln\frac{(x_2 - x_1)\sqrt{n_A}}{2s_x t(\alpha, f)}$$

where $f = n_s - 1$ with n_s being the number of determinations, from which the estimate s_x was calculated;

$\alpha=0.0388$ because for this value of α $t(\alpha,\infty) = \sqrt{2\pi e}/2$ (compare Section 2.2). The effect of the number of parallel determinations and of the accuracy of the results upon the information content is shown in the following example.

Example 3.2.1:

Al can be determined in low alloys of content 5-10 % Al as oxinate either gravimetrically (σ_x=0.025 %) or by titration of filtered oxinate (σ_x=0.035 %). In the more precise gravimetric determination provided that $n_A = 2$ parallel measurements are carried out we obtain for the information content

$$I(q||p) = \ln \frac{(10-5)\sqrt{2}}{0.025\sqrt{2\pi e}} = 4.23 \text{ nits}$$

and by titrating we obtain

$$I(q||p) = \ln \frac{(10-5)\sqrt{2}}{0.035\sqrt{2\pi e}} = 3.89 \text{ nits}$$

If we perform 4 parallel titration determinations the information gain is

$$I(q||p) = \ln \frac{(10-5)\sqrt{4}}{0.035\sqrt{2\pi e}} = 4.24 \text{ nits,}$$

therefore practically the same as in the gravimetric determination with $n_A = 2$. However, with the increasing number of parallel determinations the rate of the increase of the information content obviously diminishes (with the square root of 2): in the titration determination of Al the increase of $I(q||p)$ is 4.09-3.89 = 0.20 nits if we pass from $n_A = 2$ to $n_A = 3$ determinations, whereas in moving, e.g., from $n_A = 4$ to $n_A = 5$ determinations the information content increases only by 0.11 nits.

The effect of the number n_s of determinations, from which the estimate of the standard deviation is calculated, is traced in the next example.

Example 3.2.2:

We determine Cu in steel (0.05-0.50 %) in a photometric way with σ_x=0.006 %; for $n_A = 2$ we have

$$I(q||p) = \ln \frac{(0.5 - 0.05)\sqrt{2}}{4.1327 \times 0.006} = 3.25 \text{ nits}$$

Yet if we find s_x=0.006 % from n_s=10 determinations we have t(0.0388;9) = 2.4171 and we obtain

$$I(q||p) = \ln \frac{(0.5 - 0.05)\sqrt{2}}{2 \times 2.4171 \times 0.006} = 3.09 \text{ nits}$$

If we calculated both the mean and the sample standard deviation from the same determinations, e.g., from $n_A = n_s = 2$, we would have t(0.0388;1) = 16.3899 and I(q||p) = 1.17 nits. Thus the importance of thorough evaluation of an analytical method made from a sufficiently large number of determinations and its influence upon the information content of the results are visible.

By contrast with a distribution obtained by Bayes' theorem as in Section 3.1, this measure is not as sensitive to the inaccuracy of the a priori distribution entering the Bayes' formula (Vajda and Eckschlager, 1980). Whereas the transinformation and the conditional entropies include stochastic dependence of two random variables and we use them to investigate what information about one random variable is contained in the other, we establish, by the information gain, the similarity or the dissimilarity (the divergence) of two probability distributions.

Unless the a posteriori distribution of quantitative analysis results is unbiased, as a consequence of the rise of a systematic error or of a matrix-effect, we calculate the information gain as a difference of Kerridge-Bongard entropies in (2.3.5). We will consider results provided by this measure for some specific probability distributions. If we show a lack of a priori empirical evidence by ascribing a uniform distribution in interval $\langle x_1, x_2 \rangle$ (see Eckschlager and Štěpánek, 1979), if the analysis results are distributed normally $N(\mu, \sigma^2)$ and if they are subject to a systematic error $\delta = |X - \mu|$, where X is the true value of the concentration so that we take $N(X, \sigma^2)$ for the true distribution r(x), then

$$I(r;q,p) = \ln \frac{x_2 - x_1}{\sigma\sqrt{2\pi e}} - \frac{1}{2}\left(\frac{\delta}{\sigma}\right)^2 \qquad (3.2.1)$$

(enumerated in natural information units (nits)) (cf. Eckschlager, 1982).

The measure in (3.2.1) enables us to follow the effect of such an action that results, for instance, in lowering the mean error δ, of course in presence of simultaneously deteriorated precision, i.e., of increasing the value of σ, upon the information content of the results.

Example 3.2.3:

In the photometric determination of Mn in low-alloy steels with contents from 0.05 % through 6 % Mn we have σ_x=0.006 % and the value of a blank trial is δ=0.008 % Mn. With n_A = 2 we obtain

$$I(r;q,p) = \ln \frac{(6-0.005)\sqrt{2}}{0.006 \times 4.1327} - \frac{1}{2}\left(\frac{\sqrt{2} \times 0.008}{0.006}\right)^2 = 4.05 \text{ nits}$$

If we subtract the blank trial we diminish the mean error δ practically to zero but the standard deviation increases $\sqrt{2}$ - times so that

$$I(r;q,p) = \ln \frac{(6-0.005)\sqrt{2}}{0.006 \times 4.1327\sqrt{2}} = 5.48 \text{ nits}$$

Thus in spite of deteriorating the precision of the determination we improve the information content of the result by performing and subtracting a blank trial.

It can be easily seen that I(r;q,p) leads to the same value as the difference I(q,p) - I(r,q), which was used by Eckschlager (1979) and in Section 6.4 of our monograph. This can be easily explained in that we have started from a uniform distribution. If all three distributions are normal and comprise the true one with the mean value X and with the variance σ_r^2, the a posteriori one with μ_q and σ_q^2 and the a priori one with parameters μ_p and σ_p^2, then

$$I(r;q,p) = \ln \frac{\sigma_p}{\sigma_q} + \frac{1}{2}\left[\left(\frac{\delta_p}{\sigma_p}\right)^2 - \left(\frac{\delta_q}{\sigma_q}\right)^2 - \left(\frac{\sigma_r}{\sigma_q}\right)^2 + \left(\frac{\sigma_r}{\sigma_p}\right)^2\right] \qquad (3.2.2)$$

where $\delta_p = |X - \mu_p|$ and $\delta_q = |X - \mu_q|$. For $\mu_q = X$ and $\sigma_r = \sigma_q$ the result in (3.2.2) changes into

$$I(q||p) = \ln \frac{\sigma_p}{\sigma_q} + \frac{1}{2} [(\frac{\mu_q - \mu_p}{\sigma_p})^2 + \frac{\sigma_q^2 - \sigma_p^2}{\sigma_p^2}]$$

shown in Section 6.6 in our monograph (1979) and valid for the divergence measure of the information content, when both the a priori and the a posteriori distributions are normal.

The formula (3.2.2) makes it possible to clear up properties of I(r;q,p). It is obvious that I(r;q,p) for all entering distributions being normal will increase with δ_p, which is something like "a degree of surprise" at the result, and will decrease with increasing mean error δ. While the divergence measure I(q||p) is always non-negative I(r;q,p) can assume also negative values when completely untrue results misinform us.

Example 3.2.4:

In a photometric determination of Mn in low-alloy steels (see the preceding example) with $\sigma_q = \sigma_r = 0.008$ % we expect a content $\mu_p = 5$ % with precision $\sigma_p = 0.1$ %. Unless the result is loaded with a mean error ($\delta_q = 0$) and if we find by analysis $X = \mu_q = 4.80$ % Mn we obtain

$$I(r;q,p) = \ln \frac{0.100}{0.008} + \frac{1}{2} [(\frac{5-4.8}{0.1})^2 - 1 + (\frac{0.008}{0.1})^2] =$$

$$= 4.03 \text{ nits}$$

However, when X = 4.81 % we have $\delta_q = 0.01$ and $\delta_p = 0.20$ so that

$$I(r;q,p) = \ln \frac{0.100}{0.008} + \frac{1}{2}[(\frac{0.2}{0.1})^2 - (\frac{0.010}{0.008})^2 - 1 + (\frac{0.008}{0.1})^2] =$$

$$= 3.25 \text{ nits}$$

which is a somewhat lower information gain. If we find μ_q=4.82 % Mn then I(r;q,p) decreases to 1.90 nits and is therefore considerably lower. For μ_q=4.826 % Mn, I(r;q,p) would approximately equal zero. However, the mean error δ_q=0.026 related to σ_q=0.008 %, i.e., δ_q/σ_q=0.026/0.008= =3.25, is so large that it completely depreciates the result of the analysis.

Results, procedures, and methods of quantitative analysis need not always be judged solely according to their information content. Similarly, as Cleij and Dijkstra (1979) describe procedures of qualitative and identification analysis in terms of equivocation (the average uncertainty remaining after analysis), it is possible to qualify procedures of quantitative analysis by a posteriori inaccuracy. We will seek for an inaccuracy that is caused by the true content X not coinciding with the expected value E[ξ] of the a posteriori distribution. If we substitute in the Kerridge-Bongard measure a very narrow normal distribution with mean value X and with a negligibly small variance for the true distribution and if we integrate in (2.3.3) in a narrow interval from X-Δx through X+Δx where 2Δx << X we obtain, if p(x) is the probability density of normally distributed results,

$$H(q,p) = - \int_{X-\Delta x}^{X+\Delta x} q(x) \ln p(x) \, dx \doteq - \ln p(X) \qquad (3.2.3)$$

where p(X) is the value of p(x) at the point x = X. We employed this measure in (1980) and called its negative value A(q,p) = - H(q,p) a measure of "accuracy". Some practical uses of this "accuracy" were shown in the original paper (1980). Thus we have

$$H(q,p) = \ln \sigma\sqrt{2\pi} + \frac{1}{2}\left(\frac{x-\mu}{\sigma}\right)^2 = \frac{1}{2}[\ln 2\pi\sigma^2 + \left(\frac{\delta}{\sigma}\right)^2] \quad (3.2.4)$$

and for the measure of "accuracy"

$$A(q,p) = \frac{1}{2}[\ln \frac{1}{2\pi\sigma^2} - \left(\frac{\delta}{\sigma}\right)^2] \quad (3.2.5)$$

If determinations of several components are carried out at the same time (cf. Section 3.3) rather a measure of "mean accuracy"

$$\bar{A}(q,p) = \frac{1}{2k}\sum_{i=1}^{k} [\ln \frac{1}{2\pi\sigma_i^2} - \left(\frac{\delta_i}{\sigma_i}\right)^2] \quad (3.2.6)$$

or that of "overall accuracy"

$$A(q,p) = \frac{1}{2}\sum_{i=1}^{k} [\ln \frac{1}{2\pi\sigma_i^2} - \left(\frac{\delta_i}{\sigma_i}\right)^2] \quad (3.2.7)$$

is of interest.

The method of computation and properties of the measure of accuracy given in (3.2.5) will be discussed below.

<u>Example 3.2.5:</u>

In a photometric determination of about 5-6 % Mn in low-alloy steels (compare Example 3.2.3) we work with $\sigma = 0.008$ %. As far as a systematic error is absent, i.e., $\delta = 0$, then

$$A(q,p) = -\frac{1}{2}\ln[2\pi(0.008)^2] = 3.91 \text{ nits}$$

With an error $\delta = 0.016$, which has to be considered statistically significant as $\delta/\sigma = 2.00$, "the accuracy" decreases to the value

$$A(q,p) = -\frac{1}{2}\ln[2\pi(0.008)^2] - \frac{1}{2} \times 4.00 = 1.91 \text{ nits}$$

Thus the decrease is considerable: to 48.85 %. A titration determination of the same content in magnesium alloys is less precise; its σ is 0.025. When the results are unbiased ($\delta = 0$) we obtain $A(q,p) = 2.76$ nits, and in presence of the same error as above ($\delta = 0.016$), which, of course, does not appear as statistically significant ($\delta/\sigma = 0.64$), we get $A(q,p) = 2.56$ nits. Thus the decrease is small, only 7.25 %.

Some properties of this measure suitable also as a response function in optimizing analytical procedures have been discussed in our original paper (1980).

In judging the precision and the unbiasedness of analytical results by means of a single quantity, several characteristics have been introduced and each of them evaluates these properties to some extent only. The total error as defined by McFarren et al. (1970) and discussed by Eckschlager (1972) and Midgley (1977), is purely empirical and expresses the precision and the unbiasedness of the results rather vaguely. On the contrary, the measure of accuracy, as defined above, is derived from the assumption of a normal distribution of analytical results by characterizing the true value of the content with a narrow normal distribution around this value.

The measure A(q,p) in (3.2.4) increases with decreasing values of σ and δ, i.e., it attains higher values for more precise and unbiased results; in very unfavourable cases it can be even negative. As to the mean error δ, we shall be interested in whether it is, for a given value of σ, statistically significant and so has to be considered as a systematic error or whether it is not significant and is caused only by random nature of the measurements. Since we take into account the parameters δ and σ (and not their estimates) we can, with some simplification, set $z(\alpha) = \delta/\sigma$ and determine from the tables of the normal distribution function the corresponding significance level α, on which the particular value of δ is statistically significant with respect to σ. The dependence of A(q,p) on δ and σ can be illustrated in terms of isocurves; some of them for a few values of A(q,p) are depicted in Figure 1. It is obvious that for a given value of δ it is possible to find up to two values of σ such that the measure A(q,p) is the same. Several examples are given in our original paper (1980) demonstrating the fact that we can find, for given A(q,p) and δ and for $(\delta/\sigma) < 1$, such a value of σ that δ is not significant on the level $\alpha \geqq 0.683$ (the one sigma limit) and, for $(\delta/\sigma) > 1$, such a value of σ that δ is significant on the level $\alpha \geqq 0.683$. Accordingly, in the former case the accuracy depends mainly on σ, i.e., the term $(\delta/\sigma)^2$ is small, while in the latter case this term has quite large effect upon A(q,p).

The practical importance of measures A(q,p), $\bar{A}$(q,p), and $\tilde{A}$(q,p) consists in the following:

(1) They can serve as criteria in assessing the confidence of analytical results. In contrast to the total error they are not empirical quantities and are related to the information content of the results. However, their disadvantage is that they are not

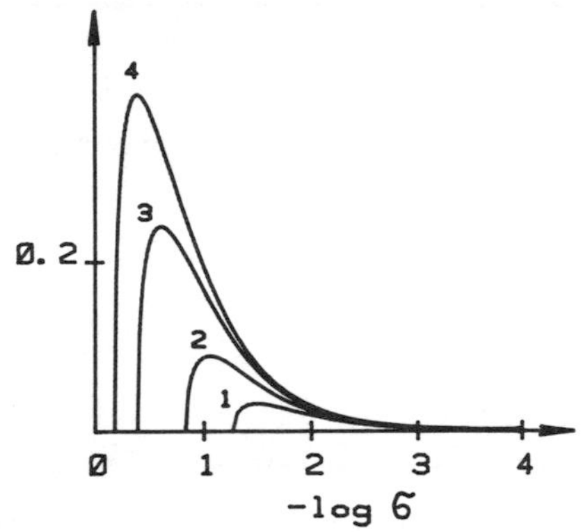

FIG. 1. Isocurves for some values of A(q,p).
Curve 1: +2; curve 2: +1; curve 3: 0; curve 4: -0.5.

relative quantities.

(2) They can be used as continuous response functions for the optimization of analytical methods, and the optimum conditions are achieved at the maximum values of these measures. If we optimize a method of determination so that the information content of the analyte of the maximum relevance be the highest, we employ one of these measures for that analyte and assign the corresponding weight to the content (cf. (4.1.2)).

In the current literature, we can find a variety of applications of various information quantities to appreciating results, methods and procedures of quantitative analysis, to assessing devices, to choosing optimum analytical methods to solve a specific analytical problem or the use of these quantities as response functions in optimizing procedures. However, no consistency exists so far in the terminology in this field.

3.3 MULTICOMPONENT QUANTITATIVE ANALYSIS

The use of information theoretic quantities and approaches in analytical chemistry has been concerned since the very beginning with instrumental methods of multicomponent analysis: let us recall, e.g., the pioneering work by H. Kaiser (1970) who introduced the so called "informing power" as the logarithm of the number of simultaneously determined components. So far a majority of papers concerned with applying information theory to analytical problems refer to multicomponent analysis, performed by means of instrumental methods.

In multicomponent analysis the input is given by a set of components $\underline{i}$ ($i=1,2, \ldots, n_o$) present in concentrations X_i; here the number n_o and the identities of individual components are either known beforehand, or they are not known and individual components have to be identified in advance (cf. Section 3.1). The output from the first subsystem is presented by more or less mutually overlapping signals that have intensities y_{ij} ($j=1,2, \ldots, m$; $m \geqq n_o$) in position Z_j. For the results ξ_i its basic validity is the same as in the one-component analysis; in addition the unfavourable effect of mutual overlapping appears if we have the uncertainty of the results in mind. In multicomponent quantitative analysis we consider as a rule the simplest relationship between concentration and intensity of the signal for both the calibration and the analytical functions

$$E[\eta_i] = S_{ij} X_i$$

$$E[\xi_i] = \frac{1}{S_{ij}} E[\eta_i]$$

where S_{ij} (= a constant) is the partial sensitivity of

the determination of the component $\underline{i}$ according to the signal intensity η_i in position Z_j. We try to work under such conditions that these relationships be at least approximately fulfilled. The input-output relation is given by the matrix of partial sensitivities S_{ij}; for a perfectly selective determination of all components we need to have large S_{ii} and $S_{ij} = 0$ $(i \neq j)$ at the same time (or S_{ij} is at least small enough compared to S_{ii}). H.Kaiser (1972) built his definition of specificity and selectivity on partial sensitivities. Partial sensitivities S_{ij} can assume relatively high values, in which they differ from elements of matrix a_{ji} treated in Section 3.1 where $0 \leqq a_{ji} \leqq 1$, which made it possible to substitute them in formulae for uncertainties. Therefore we sometimes take relative partial sensitivities into account

$$d_{ij} = \frac{S_{ij}}{\sum\limits_{i} S_{ij}} \qquad j=1,2, \ldots, m \tag{3.3.1}$$

so that $0 \leqq d_{ij} \leqq 1$. Then we can qualify the uncertainty dependent on the selectivity of multicomponent qualitative or quantitative analysis by Shannon's entropy

$$H_j = -\sum_{i=1}^{n} h_{ij} \operatorname{ld} h_{ij} \qquad j=1,2, \ldots, m \tag{3.3.2}$$

where h_{ij} equals a_{ji} or d_{ij} and $0 \leqq h_{ij} \leqq 1$. A still better quantity characterizing the selectivity is the relative entropy (Eckschlager, 1982) which reflects the uncertainty in determining various components $\underline{i}$ if measurements are made in position Z_j $(j=1,2, \ldots, m)$. We will return to this quantity in greater detail in Section 4.2; here we

$$H_{r(j)} = \frac{H_j}{\mathrm{ld}\ n} \quad j=1,2,\ \ldots,\ m \qquad (3.3.3)$$

are going to show the way of computing H_j and $H_{r(j)}$.

Example 3.3.1:

For determination of the o-, m-, and p-cymens in a technical product by IR-spectrometric method the following absorbencies for three different values Z_j are given (compare Doerffel and Eckschlager (1981), p.58) :

X_i	Z_j in cm^{-1}		
	758	785	818
o-cymen	209.60	2.17	0.65
m-cymen	1.07	166.50	1.07
p-cymen	0.17	1.19	173.80
	210.84	169.86	175.52

First we find values d_{ij}, for instance for $Z_2 = 785\ cm^{-1}$, in this way:

$$d_{12} = 2.17 / 169.86 = 0.012775$$
$$d_{22} = 166.50 / 169.86 = 0.980219$$
$$d_{32} = 1.19 / 169.86 = 0.007006$$

$$\sum_{i=1}^{3} d_{i2} = 1$$

In a similar way we determine the values d_{ij} for the other j's and we obtain the following matrix:

	758 cm^{-1}	785 cm^{-1}	818 cm^{-1}
o-cymen	0.994118	0.012775	0.003703
m-cymen	0.005076	0.980219	0.006096
p-cymen	0.000806	0.007006	0.990201

Afterwards we can compute entropies H_j for individual positions Z_j as

$$H_j = - \sum_{i=1}^{3} d_{ij} \text{ ld } d_{ij}$$

and

$$H_{r(j)} = \frac{H_j}{\text{ld } 3}$$

and we obtain these values:

Z_j	H_j	$H_{r(j)}$
758 cm^{-1}	0.05543	0.03497
785 cm^{-1}	0.15876	0.10017
818 cm^{-1}	0.08883	0.05605

The uncertainties of the results of measurements carried out in the particular positions Z_j (j=1,2,3) enumerated by the values H_j or $H_{r(j)}$ are relatively small.

The quantity

$$H_r = -\frac{\sum_{i=1}^{n}\sum_{j=1}^{m} h_{ij} \text{ ld } h_{ij}}{m \text{ ld } n} \qquad (3.3.4)$$

is characteristic for the selectivity of the entire procedure. It holds that $0 \leqq H_r, H_{r(j)} \leqq 1$. The calculation of H_r according to (3.3.4) is analogous to that of $H_{r(j)}$.

<u>Example 3.3.2:</u>

If we utilize the values H_j and $H_{r(j)}$ from the preceding example we obtain

$$\sum_{j=1}^{3} H_j = 0.30302 \qquad \sum_{j=1}^{3} H_{r(j)} = 0.19119$$

Thus the quantity H_r characterizing the determination of the o-, m-, and p-cymens turns out

$$H_r = \frac{1}{3 \times \text{ld } 3} \sum_{j=1}^{3} H_j = 0.06373$$

For a perfectly selective procedure in multicomponent analysis we would obtain $H_r = 0$; an entirely non-selective procedure yields $H_r = 1$.

Here we have to stress that d_{ij} are neither probabilities nor relative frequencies in contrast to a_{ji}. They are relative values of an operator, ascribing the signal intensity in position Z_j to the concentration X_i of the component i. This operator is usually taken as constant, or we work in a domain in which this condition is satisfied at least approximately.

Another component of the a posteriori uncertainty is that due to both the random and systematic errors of the result, as was discussed in the preceding section.

In customary analytical practice, two marginal problems of applying multicomponent analysis exist:

(i) Precise determination of a few components when we know in advance what components and in what approximate contents are present in the analysed sample.

(ii) Semiquantitative (often only as to the order of magnitude) determination of the contents of all components in the sample, about the composition of which we know practically nothing, i.e., a screening.

There exist of course a variety of possibilities between both extremes. We summarize the amout of information gained by performing a multicomponent quantitative analysis as

$$M(q||p) = \sum_{i=1}^{n} I_i(q||p) \qquad (3.3.5a)$$

or

$$M(r;q,p) = \sum_{i=1}^{n} I_i(r;q,p) \qquad (3.3.5b)$$

where $I_i(q||p)$ and $I_i(r;q,p)$ are the divergence measures from (2.3.2) and (2.3.5) for the ith component. In the most frequent case of a uniform a priori distribution and a normal distribution of analysis results, $I(q||p)$

depends on the precision of the measurements (cf. Section 6.3 of our monograph, 1979), and I(r;q,p) by (3.2.1) depends on both the precision and the unbiasedness of the results. Then in case (i) we gain a great amount of information by high precision and by unbiasedness of the determinations. In the case of a screening, information is gained by a great number of components, the presence of which can be excluded or the content of which in the sample we estimate at least in the order of magnitude. H. Kaiser (1970) obviously had in mind screening when he introduced the quantity "informing power", depending only on the number of simultaneously determined components regardless of the precision of the results. The dependence of the information content on the precision is taken into account later by adopting the divergence measure. If a screening is followed by a precise determination of individual components the formula for two normal distributions applies (cf. Section 6.6 of our monograph)

$$I(q||p) = \ln \frac{\sigma_p}{\sigma_q} + \frac{1}{2}\,\frac{(\mu_q - \mu_p)^2 + \sigma_q^2 - \sigma_p^2}{\sigma_p^2}$$

or the formula in (3.2.2) is of use.

Besides selectivity, precision, and unbiasedness another potential source of a posteriori uncertainty can be sought for in the random nature of a signal position; its importance appears mainly when the divergence measure is adopted. Let us assume again a uniform distribution $U(x_1,x_2)$ when no prior empirical evidence concerning X was collected ($U(x_1,x_2)$ denotes the uniform distribution with the probability density $p(x) = (x_2-x_1)^{-1}$ for $x \in \langle x_1,x_2 \rangle$, $p(x) = 0$ otherwise). If the fixed unknown content X can be expected to lie in the middle of the interval $\langle x_1,x_2 \rangle$ we can consider the a priori distribution

as normal $N(\mu_p(x), \sigma_p^2(x))$ where $\mu_p(x) = (x_1 + x_2)/2$ and $\sigma_p(x) = (x_2 - x_1)/K$ with $K \geqq 6$. Since the results of quantitative instrumental analyses are usually normally distributed $N(\mu_q(x), \sigma_q^2(x))$ the information content takes on the value, in the case of the uniform a priori distribution and provided that
$x_1 + 3\sigma_q(x) \leqq \mu_q(x) \leqq x_2 - 3\sigma_q(x)$,

$$I(q||p) = \ln \frac{(x_2 - x_1)\sqrt{n_p}}{\sigma_q(x)\sqrt{2\pi e}} \qquad (3.3.6)$$

where n_p is the number of parallel determinations (cf. Section 6.3 of our monograph). If the a priori distribution is that normal $N(\mu_p(x), \sigma_p^2(x))$ above and $\sigma_q(x) \ll x_2 - x_1$ and $\mu_p(x) \approx \mu_q(x)$, which is common in analytical practice, the amount of information

$$I(q||p) = \ln \frac{(x_2 - x_1)\sqrt{n_p}}{\sigma_q(x) K \sqrt{e}} \qquad (3.3.7)$$

where $K \geqq 6$.

The formula in (3.3.7) can be rewritten as

$$I(q||p) = C + 2.36 \log \frac{1}{\sigma_q} \quad \text{where} \quad C = \ln \frac{(x_2 - x_1)\sqrt{n_p}}{K\sqrt{e}}$$

depends only on the a priori probability distribution and on the number of parallel determinations carried out. If we introduce $p\sigma_q = -\log \sigma_q$ the information gain maps as linear dependent on $p\sigma$ in graphic representation. The quantity $p\sigma$ will take place later in our considerations of the dependence of the information relevance on the information content; yet its importance is also in the fact that the size of the standard deviation varies in the range of several orders of magnitude for analytically controllable concentrations from 10^{-12} through 10^{2} % and thus the logarithmic scale is more suitable to work with its values.

In the situation when the results of a multicomponent instrumental analysis do not refer to the value x_i, but rather to intensity y_i and to position z_j of the signal, the information content appears as

$$I(q||p) = \ln \frac{(y_{max} - y_{min})\sqrt{n_p}}{\sigma_q(y)K\sqrt{e}} \qquad (3.3.8)$$

with $K = \sqrt{2\pi}$ for the uniform, and $K \geqq 6$ for the normal, a priori distribution. This formula is analogous to (3.3.6) and (3.3.7) but, whereas the values x_1, x_2, and $\sigma_p(x)$ refer to our preliminary knowledge of the composition of the sample to be analysed, y_{max}, y_{min}, and $\sigma_q(y)$ are given by technical parameters of the analytical instrument. If both the intensity η_i and the position ζ_j of the signal are sources of relevant information and they are mutually independent, we define the information content as

$$I(q||p) = \int_{y_{min}}^{y_{max}} q(y) \ln \frac{q(y)}{p(y)} dy +$$

$$+ \int_{z_{min}}^{z_{max}} q(z) \ln \frac{q(z)}{p(z)} dz \qquad (3.3.9)$$

In the case of uniform or normal a priori distributions

p(y) and p(z), and normal distributions q(y) and q(z) of the results, the formula in (3.3.9) yields

$$I(q||p) = \ln \frac{(y_{max} - y_{min})(z_{max} - z_{min}) n_p}{\sigma_q(y)\,\sigma_q(z)\,K^2 e} \qquad (3.3.10)$$

where $K^2 = 2\pi$ for the uniform, and $K^2 \geqq 36$ for the normal a priori distribution.

If several components are determined simultaneously, and if both the signal position and the intensity are sources of relevant information, then the total amount of information

$$M(q||p) = \sum_{i=1}^{n} I_i(q||p)$$

where we sum values from (3.3.10) obtained for individual components (similarly as in (3.3.4)).

A relatively complicated factor is the effect of sensitivity S_{ij} upon the amount of information obtained from the results of a multicomponent quantitative analysis. Besides the influence upon selectivity as described at the beginning of this section, the value of S_{ij} also affects the precision of the determination. The variance of the results is connected with the variance of the signal intensity measurement in the case of a linear calibration or analytical function by the relation

$$\sigma(x) = \frac{1}{S_{ij}}\,\sigma(y) \qquad (3.3.11)$$

so that the measure of inaccuracy of the results ξ_i

depends on the variance of the signal intensity measurements $\sigma^2(y)$ and on the sensitivity S_{ij} according to (3.2.4)

$$H(q,p) = \frac{1}{2} [\ln 2\pi\sigma^2(y_i) + (S_{ij} \frac{\delta}{\sigma(y_i)})^2] - \ln S_{ij}$$

$$j=1,2, \ldots, m \qquad (3.3.12)$$

If we calculate the amount of information by (3.3.6), (3.3.7) or (3.3.8) the connection of ranges between x_1 and x_2 and between y_{min} and y_{max} takes its place. It is obvious that generally x_1 need not correspond to y_{min} and x_2 to y_{max}; if we work with a sample the composition of which we do not know, the interval $\langle x_1, x_2\rangle$ will be wide, e.g., $\langle 0, 100\%\rangle$ and it can happen that it will be wider for some given sensitivity than that which can be covered by values from y_{min} through y_{max} (Figure 2). The relationships between cases when we start from a priori distributions p(x) and p(y) respectively have been discussed in detail (Eckschlager, 1976, 1981).

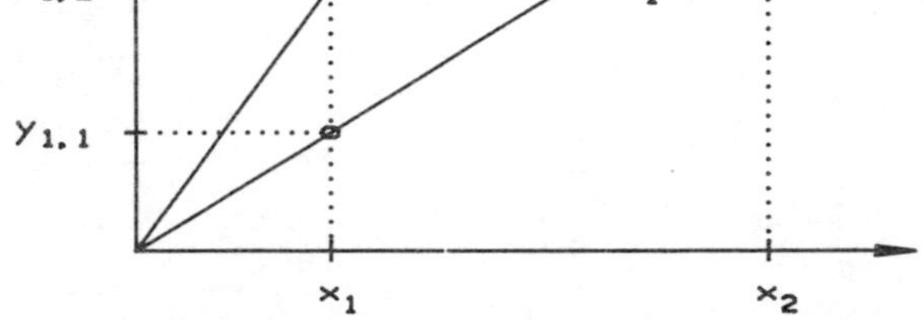

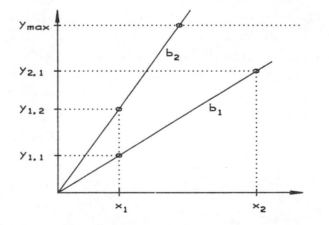

FIG. 2. Linear dependence y=bx where b=dy/dx = const. in the whole interval $\langle x_1, x_2\rangle$.

We sometimes cannot reach perfect discrimination of neighbouring signals. In the most simple case of two components, when we measure intensities in two positions z_j (j=1,2), we have a system of equations

$$y_1 = s_{11}x_1 + s_{21}x_2 \qquad s_{11} > s_{21}$$
$$y_2 = s_{12}x_1 + s_{22}x_2 \qquad s_{22} > s_{12} \qquad (3.3.13)$$

All the partial sensitivities s_{ij} can be determined in experiments with pure materials. Let us denote

$$w_1 = \frac{s_{21}}{s_{11}} < 1 \quad , \quad w_2 = \frac{s_{12}}{s_{22}} < 1 \qquad (3.3.14)$$

as measures of overlapping of the signals. If we carried out the determination of both components regardless of overlapping of the signals, a systematic error $\delta_1 = w_1x_2$ would arise in determining the first component in position 1 and an error $\delta_2 = w_2x_1$ in determining the second component in position 2. If we handle measured values in the same way as those found by solving the system of equations (3.3.13) we remove systematic errors, but the variance of either result will increase. In the simple situation ($s_{11} \approx s_{22}$, $s_{12} \approx s_{21}$, $y_1 \approx y_2$) we find that

$$\sigma(x_1) = \frac{\sigma(y)}{s_{11}} \frac{\sqrt{1 + w_1^2}}{1 - w_1^2}$$

$$(3.3.15)$$

$$\sigma(x_2) = \frac{\sigma(y)}{s_{22}} \frac{\sqrt{1 + w_2^2}}{1 - w_2^2}$$

Apparently the information content given by (3.3.6) or (3.3.7), or the a posteriori uncertainty, will be almost the same, for small values of w_1 and w_2, whether we determine both components with regard to overlapping signals or regardless of them. It can be shown as follows.

Example 3.3.3:

These are the values of absorbencies for the determination of m- and p-cymens (see Example 3.3.1):

	785 cm^{-1}	818 cm^{-1}
m-cymen	166.50	1.07
p-cymen	1.19	173.80

w_1 = 1.19/166.50 = 0.00713 w_2 = 1.07/173.80 = 0.00616

The values $\sqrt{\dfrac{1 + w_1^2}{1 - w_1^2}} = 1.00008$ and $\sqrt{\dfrac{1 + w_2^2}{1 - w_2^2}} = 1.00006$

are so close to one that their effect is practically of no use.

When w_1 and w_2 are large enough, and especially for small $\sigma(y)$ and high values of S_{ii}, the information content increases considerably (or the uncertainty after analysis decreases) provided that we process the results in such a way that we take into account overlapping of the signals.

In multicomponent quantitative analysis we often obtain, without the possibility of avoiding it, signals of which the position and the intensity contain irrelevant information in themselves. We appreciate the relevance of specific information according to how it contributes to the resolution of a given problem, e.g., in technical or economic decision making, in verifying hypotheses, in determining diagnosis, etc. Relevance is usually understood qualitatively, apparently because of the difficulty of quantification of this concept. A specific, even if rather formal and imperfect, attempt at quantifying relevance can be made by the following reasoning: a decision on whether specific information gained from analysis is relevant or irrelevant can be described as inserting it into a subset of relevant or irrelevant information. If the subset of relevant information is sharp for information $i \in I$, i.e., $I_r \subset I$, the appurtenance of $\underline{i}$ to I_r is given by mapping

$$\varphi(i) = \begin{cases} 0 & i \notin I_r \\ 1 & i \in I_r \end{cases}$$

and we have only two alternatives: the information is completely relevant or completely irrelevant. Yet if the subset of relevant information is fuzzy (Zadeh 1965, 1968), i.e., $I_r \lessapprox I$, the appurtenance of $\underline{i}$ to I_r is described by mapping $\varphi(i)$ so that $0 \leqq \varphi(i) \leqq 1$, i.e., continuously assuming values from zero to one and being the closer to one the more information $\underline{i}$ belongs to the fuzzy subset I_r. Thus we can characterize the relevance by a relevance coefficient $k(r)$ having meaning of a weight and being understood as the value of the function $\varphi(i)$ for specific information.

The amount of exploitable information that we obtain from a multicomponent analysis can be expressed as

$$M(q||p)_E = \sum_{i=1}^{n} k_i(r)\ I_i(q||p)$$

where $k_i(r)$ is the relevance coefficient of information about the i-th component ($i=1,2, \ldots, n$). As far as the analysis result is employed as information for decision making, for instance in analytical quality control, in medical diagnosis, in deposit prospecting, etc., we can take $k_i(r)$ for constant (for a given component). In carrying out analyses, the results of which are to serve to the corroboration or to the denial of a scientific hypothesis or theory, we often find out that the relevance of particular information is, at least to some extent, dependent on its content or, in the case of information obtained from measurements, on the accuracy of the result.

The course of the change of the relevance coefficient $k(r)$ with the accuracy $\underline{t}$ can be described by the differential equation

$$\frac{dk}{dt} = f(k)$$

If this rate of the change is anticipated as constant in the range $t \in \langle t_1, t_2 \rangle$, i.e.,

$$\frac{dk}{dt} = a, \quad a > 0 \ \text{const.}$$

the solution for the initial condition $t = t_1$, $k = 0$ appears as

$$k = \begin{cases} 0 & \text{for } t < t_1 \\ a(t - t_1) & \text{for } t_1 \leqq t \leqq t_2 \\ k_{max} & \text{for } t > t_2 \end{cases}$$

It is, of course, possible to assume other kinds of

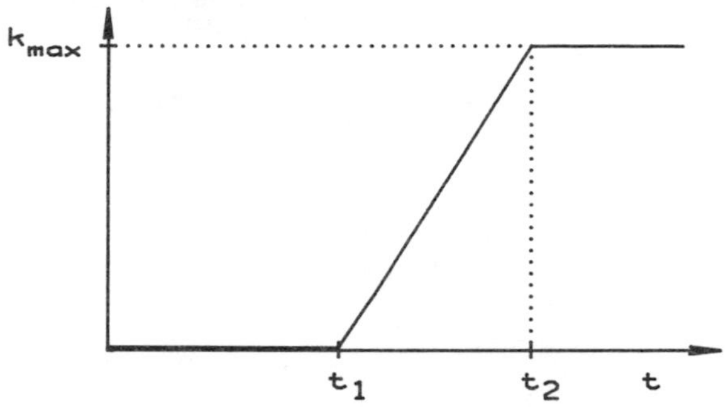

FIG. 3. A linear dependence of the relevance coefficient on the accuracy of the result.

dependence of the relevance on the information content or on the accuracy, e.g.,

$$\frac{dk}{dt} = k(a - bk) \qquad 0 < a \leqq b$$

Then the solution is the Robertson law of growth, i.e., the function

$$k = \frac{k_{max}}{1 + C\,\exp(-at)}$$

where $k_{max} = a/b$. From the initial condition $t = t_o$, $k = k_{max}/2$ we determine $C = \exp(at_o)$ so that the dependence sought-for is

$$k = \frac{a}{b}\,\frac{1}{1 + \exp[-a(t-t_o)]}$$

The slope of the curve in the inflection point $(t_o, a/2b)$ has the size $a^2/4b$.

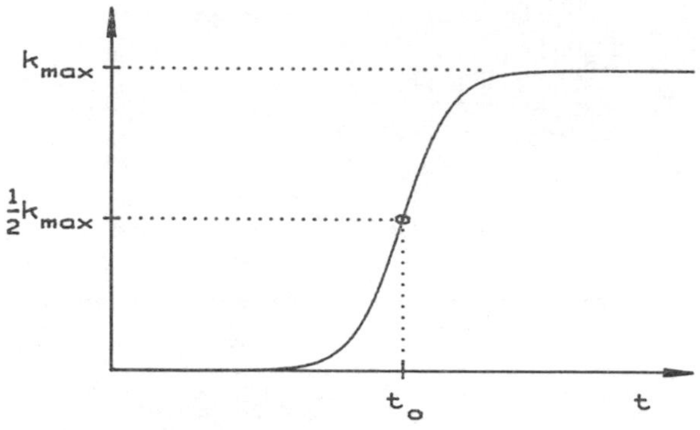

FIG. 4. The Robertson law of growth of the relevance coefficient.

Thus, e.g., if $k_{max} = a/b = 0.75$ and we claim $k=0.05\ k_{max}$ for $t_1 = 3.3$ and $k = 0.95\ k_{max}$ for $t_2 = 4.0$ we substitute $t_o = 3.65$ so that

$$k = \frac{0.75}{1 + \exp[-a(t - 3.65)]}$$

and we obtain the value of a from the conditions in the boundary limits of the accuracy t_1 or t_2.

The value k_{max} is, of course, attached, as for any weight, on the basis of experience (i.e., subjectively) with regard to the hypothesis to be verified, to the degree to which given information facilitates a decision, etc.

Example 3.3.4:

In GLC analysis (n_A=6) of plant oils by a determination

TABLE 3. The amount of information from a multicomponent analysis when relevance coefficients are constant.

acid	$k_i(r)$	$\langle x_1, x_2 \rangle$	s	$\bar{x} \pm ts/\sqrt{n_A}$	$I_i(q\|\|p)$	$k_i(r)I_i(q\|\|p)$
C_{16}: 0	0.50	2- 6	0.0858	5.24±0.09	3.076	1.538
C_{16}: 1	0.10	0- 1	0.0381	0.30±0.04	2.501	0.250
C_{18}: 0	0.85	0- 5	0.0857	1.95±0.09	3.300	2.805
C_{18}: 1 t	0.90	0- 3	0.0476	1.62±0.05	3.377	3.039
C_{18}: 1 c	0.85	30-66	0.3049	56.12±0.32	4.005	3.404
C_{18}: 2 ct	0.80	0- 3	0.0667	0.58±0.07	3.040	2.432
C_{18}: 2 cc	0.95	10-36	0.2002	18.72±0.21	4.100	3.895
C_{18}: 3	0.95	5-10	0.1811	8.10±0.19	2.552	2.424
C_{20}: 0	0.10	0-1.5	0.0858	0.71±0.09	2.095	0.209
C_{20}: 1	0.10	0- 5	0.1429	3.17±0.15	2.789	0.279
C_{22}: 0	0.10	0- 1	0.0286	0.21±0.03	2.788	0.279
C_{22}: 1 t	0.85	0- 5	0.1810	3.35±0.19	2.553	2.170
				Sum:	36.176	22.725

of methylesters of individual carboxyl acids it is possible to attach relevance coefficients $0.05 \leqq k_i(r) \leqq 0.95$ to the acids according to how they affect taste, physiological, and technological qualities of the vegetable oil and afterwards to use them as constant weights. If we indicate individual acids with the number of atoms C in a molecule and with the number of double bonds, in which we discriminate cis(c) and trans(t) forms of the isomers, we can make up the Table 3. Thus $M(q||p) = 36.176$ nits and $M(q||p)_E = 22.725$ nits. Apparently the exploitable amount of information is smaller, which is due to low relevance of a few acids.

Example 3.3.5:

For the determination of a content from 0.05 through 0.50 % Cu in steel we need such an analytical method that would enable us to discriminate contents of Cu with $\Delta = 0.015$ till 0.020 %. Information making possible a discrimination with $\Delta = 0.020$ % is assigned a coefficient $k(r) = 0.05$ and we provide more precise information discriminating with 0.015 % with a relevance coefficient $k(r) = 0.95$. If we carry out two parallel determinations the value $\Delta = 2 \times 1.96 \times \sigma / \sqrt{2} = 2.772\sigma$ refers to the discrimination with probability 0.95. It means that we obtain $\sigma = 0.005411$ for $\Delta = 0.015$, i.e., $p\sigma = -\log \sigma = 2.2667$ whereas we have $\sigma = 0.007215$ and $p\sigma = 2.1418$ if $\Delta = 0.020$. For the Robertson function we obtain

$$k = [1 + \exp\{-47.14(p\sigma - p\sigma_o)\}]^{-1}$$

with $p\sigma_o = 2.2042$. If we adopt a photometric method, for which $\sigma = 0.006$, it results in

$$I(q||p) = \ln \frac{(0.50 - 0.05)\sqrt{2}}{0.006\sqrt{2\pi e}} = 3.245 \text{ nits.}$$

However, the method does not provide fully relevant results; for $\sigma = 0.006$ we obtain $k(r) = 0.697$ and we will substitute $k(r)I(q||p) = 0.697 \times 3.245$ in computing the exploitable amount of information obtained by analysing Cu in steel.

Literature dealing with applications of information theory in multicomponent and instrumental quantitative analysis has become extensive. The state-of-the-art by the end of 1978 is set out in our monograph (1979) and from recent papers we call attention to (Liteanu and Rica, 1979; Eckschlager and Štěpánek, 1982; Frank et al., 1982).

3.4 TRACE ANALYSIS

The practical importance of trace analysis increases constantly, as requirements for determining smaller amounts of different substances in materials often quite composite grow in importance, although we cannot always collect arbitrary great amounts of the sample (e.g., clinical biochemistry). The a posteriori uncertainty is enhanced by factors not found in other kinds of analyses - or at least not to such an extent - e.g., sample contamination, losses in handling, the purity of chemicals and of water, etc.

The input and the input-output relation in trace analysis are the same as in quantitative analysis but in the output a signal of weak intensity arises with a noise background which may be distributed normally, or have a truncated normal distribution, or a log-normal distribution. The signal-to-noise-ratio is an important feature for judging a signal. The information content has been even equalled to the logarithm of this ratio; that simplification is, of course, hardly acceptable. In appreciating the results and methods of trace analysis the divergence measure $I(q||p)$ is of major importance or we can use $I(r;q,p)$ according to (2.3.5). The measure of "accuracy" can also be usefully adopted here.

As a rule we will use a continuous uniform distribution $U(0,x_1)$ for the a priori one where x_1 is the highest anticipated content of a trace component to be determined in the analysed sample. In order to express the uncertainty after analysis, or the amount of information, we have to distinguish two cases in connection with an a posteriori distribution.

(i) The content of the component to be determined is less than the determination limit, i.e., $X_i < x_o$ and the result of analysis consists in finding out that situation.

(ii) The content of the component to be determined is above the determination limit, i.e., $X_i \geqq x_o$, and the content of the trace component can be determined quantitatively. Then the uncertainty and the information content depend on the distribution of signal intensities and therefore on that of the results.

Both cases have been treated in an early paper employing the divergence measure to analytical needs (Eckschlager, 1975) and later work has employed them more exactly (Eckschlager and Štěpánek, 1978, 1981; Štěpánek and Eckschlager, 1979).

If the true content of the trace component $X_i \leqq x_o$ we express our uncertainties by an a priori uniform distribution $U(0,x_1)$ and by an a posteriori one $U(0,x_o)$. Hence the uncertainty after analysis $H(q) = \ln x_o$ and

$I(q||p) = \ln (x_1/x_o)$ is the amount of information ($x_1 > x_o$). Even in this case the information content of such a finding can be great enough, when the upper limit of the contents is anticipated larger by a few orders of magnitude than is the determination limit of the method adopted.

Example 3.4.1:

Already for the case when $x_1 = 25\ x_o$ we obtain $I(q||p) = \ln 25 = 3.22$ nits so that the finding that the content of the analyte in the sample is under x_o brings approximately the same information gain as does the photometric determination of Cu in steel (compare Example 3.3.5). In addition such a statement can be of high relevance, for instance, in analyses of substances with high purity.

In the alternative case where $X_i > x_o$, the signal can be subject either to a log-normal distribution or to a normal one. Most frequently we encounter a shifted log--normal distribution commencing from the determination limit with parameter $\mu = \ln kx_o$ (as in our paper, 1978), i.e., the distribution

$$q(x) = \begin{cases} [(x-x_o)\sigma\sqrt{2\pi}]^{-1}. \\ .\exp\left\{-\frac{1}{2}\left[\frac{\ln(x-x_o) - \ln kx_o}{\sigma}\right]^2\right\} & \text{for } x > x_o \\ 0 & \text{for } x \leqq x_o \end{cases}$$

with expected value $E[\xi] = x_o(1 + k.\exp[\sigma^2/2])$ and variance $V[\xi] = kx_o.\exp[\sigma^2].(\exp[\sigma^2] - 1)$. The information gain differs from that in the former case by the term $\ln(\sqrt{n_p}/k\sigma\sqrt{2\pi e})$ where n_p is the number of parallel determinations. If the expected value of the shifted log-normal distribution is close to x_o, i.e., if $\underline{k}$ is sufficiently small and the distribution is strongly asymmetric (the common case), the term is positive, so that the amount of information is greater than that in the case when the value of contents is less than the determination limit.

If the results of a trace analysis follow a truncated Gaussian distribution

$$q(x) = \begin{cases} \left\{[1-\Phi(z_o)]\sigma\sqrt{2\pi}\right\}^{-1} \exp[-\frac{1}{2}(\frac{x-\mu}{\sigma})^2] & \text{for } x > x_o \\ 0 & \text{for } x \leqq x_o \end{cases}$$

(because the mean value is close to the determination limit, i.e., $x_o \leqq \mu \leqq x_o + 3\sigma$) where $z_o = (x_o - \mu)/\sigma$ and $\Phi(z)$ is the normal distribution function, the information gain depends on the deviation of μ from the determination limit by the formula

$$I(q||p) = \ln \frac{x_1}{\sigma\sqrt{2\pi e}} + \frac{1}{2}\,\frac{z_o\varphi(z_o)}{1-\Phi(z_o)} - \ln[1-\Phi(z_o)] \qquad (3.4.1)$$

($\varphi(z)$ is the normal frequency function).

The dependence of $I(q||p)$ from (3.4.1) on μ can be shown in the following example.

<u>Example 3.4.2:</u>

For a trace analysis when we carry out $n_A = 3$ parallel determinations by a method working with $\sigma = 2.5 \times 10^{-5}$ and when we expect, from experience, the maximum content of the analyte $x_1 = 10^{-3}$ %, the information content for

$\mu >> x_o$ turns out to be

$$I(q||p) = \ln \frac{10^{-3}\sqrt{3}}{2.5 \times 10^{-5}\sqrt{2\pi e}} = 2.819 \text{ nits}$$

For a case when $\mu = x_o + 0.5\sigma$, i.e., $z_o = 0.5$ (Figure 5a), we obtain

$$I(q||p) = 2.819 + \frac{1}{2}\,\frac{0.5 \times 0.3521}{0.6915} - \ln 0.6915 = 3.315 \text{ nits;}$$

if $\mu = x_o$, i.e., $z_o = 0$ (Figure 5b), we have

$$I(q||p) = 2.819 - \ln 0.5 = 3.512 \text{ nits}$$

whereas for $\mu = x_o - 0.5\sigma$ and $z_o = -0.5$ (Figure 5c) we get

$$I(q||p) = 2.819 - \frac{1}{2}\,\frac{0.5 \times 0.3521}{0.3085} - \ln 0.3085 = 3.709 \text{ nits.}$$

The found values can be compared with the course of curve N in Figure 6.

With increasing content of the sought-for component the log-normal distribution becomes symmetric and the truncation of the normal distribution decreases. Then the information gains converge, in either case, to a common limit

$$I(q||p) = \ln \frac{x_1}{\sigma\sqrt{2\pi e}} \tag{3.4.2}$$

valid in the case when the a posteriori distribution is

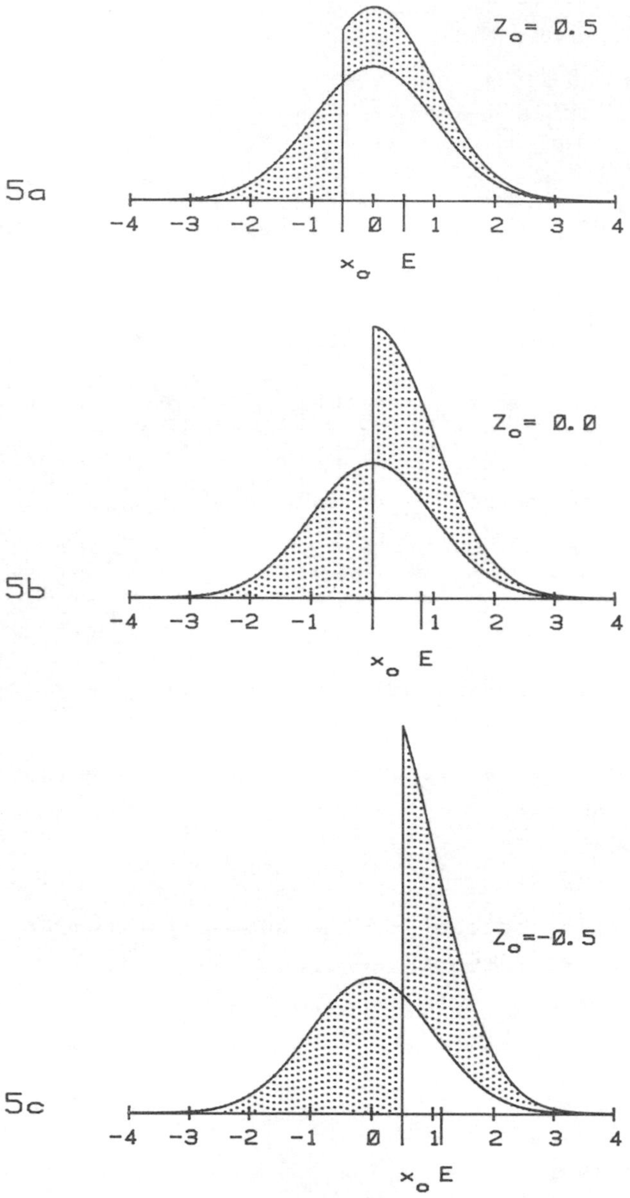

FIG. 5. Normal truncated distribution (see Example 3.4.2).

normal $N(\mu, \sigma^2)$. The dependence of the information gain on the content of the trace component is illustrated in Figure 6.

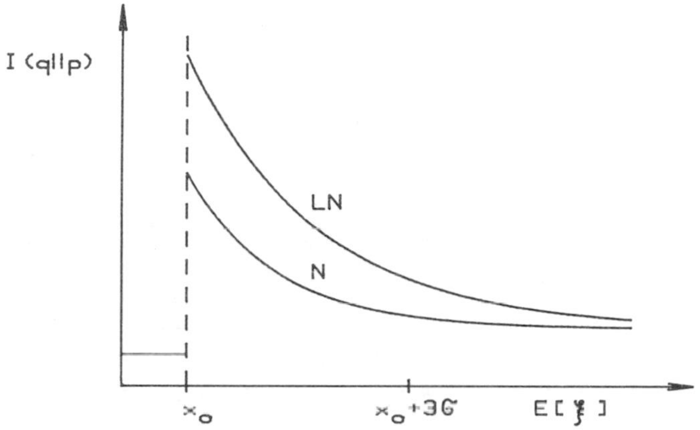

FIG. 6. The dependence of I(q||p) on the content of the component to be determined. I(q||p) is constant for $X < x_0$. Curve N belongs to the normal distribution (truncated) and curve LN to the log-normal distribution. Both tend to the same limit given in (3.4.2).

In addition to the papers above quoted, the information theory approach to local analysis (a special case of trace analysis) was examined by Danzer (1974). Liteanu and Rica (1979) elaborated lucidly the probabilistic theory of trace analysis and they paid attention also to the information quantities of the results in terms of communication theory. Nevertheless all the possibilities which information theory provides, for passing judgments on devices and on ways of processing analytical signals for the needs of trace analysis, have not been so far utilized. Trace analysis is not, of course, a homogeneous discipline in the problems dealt with: for instance, in output analysis of high purity materials we basically deal with identification analysis of the screening type, while in controlling the purity we require a precise and accurate determination of the contents of a few ingredients given beforehand. In studying the effectiveness of a purification process, we postulate that sensitive recording of the changes of concentration of one or several ingredients is possible, in which the unbiasedness is not decisive (we often do not require the knowledge of the real concentrations of the ingredients). In trace analyses of biological materials, performed mostly on a micro- or semimicro scale in clinical biochemistry where the results serve for diagnostic purposes, we require precision and accuracy as great as possible if the uncertainty of the diagnosis is to be a minimum; after all, the so called "normal physiological" values are often affected by the adopted method of determination. The grounds for applying information theory in clinical biochemistry examinations have been presented by Büttner (1982). Different again are the requirements laid upon analysis results in investigating environmental pollution levels, etc. All these problems, very topical today, can be qualified according to uncertainty or to information content and to relevance of information provided by trace analysis.

Similarly as in Section 3.2 we can approach trace analyses in such a manner that we appreciate the accuracy of the supposed distributions of the reasults. We will characterize again the true value of the trace content by a very narrow normal distribution $N(X, \sigma_q^2)$ with a small value of σ_q^2. Then the inaccuracy can be measured by the measure given in (3.2.4) provided that a normal distribution governs the results (the probability density at the value of the determination limit must be practically equal to zero). In the more realistic situation when the results are governed by a truncated normal distribution, arising from the preceding one, the following measure of inaccuracy can be derived

$$H(q,p) = \ln \sigma\sqrt{2\pi}\,[1 - \Phi(z_o)] + \frac{1}{2}\left(\frac{x - \mu}{\sigma}\right)^2 \qquad (3.4.3)$$

(we see immediately that formula (3.2.4) follows from it when z_o tends to minus infinity). Here, of course, the parameter μ is not the expected value of the truncated normal variable; the expectation

$$E[\xi] = \mu + \sigma \frac{\varphi(z_o)}{1 - \Phi(z_o)}$$

(resulting in μ when the truncation disappears).

Example 3.4.3:

The calculation of the expectations for data from the preceding example yields:
for $\mu = x_o + 0.5\sigma$ we obtain

$$E[\xi] = x_o + 0.5\sigma + \frac{0.3521}{0.6915}\sigma = x_o + 1.009\sigma;$$

for $\mu = x_o$ we obtain

$$E[\xi] = x_o + \frac{0.3989}{0.5}\sigma = x_o + 0.798\sigma;$$

for $\mu = x_o - 0.5\sigma$ we obtain

$$E[\xi] = x_o - 0.5\sigma + \frac{0.3521}{0.3085}\sigma = x_o + 0.641\sigma.$$

These values are marked in Figures 5a, 5b, and 5c. The calculation of them is of importance also for considerations concerning the detection and determination limits (compare Section 4.4).

If we denote, as earlier, the deviation of the expected value from the true value of the content, i.e., $\delta = X - E[\xi]$, and insert it into (3.4.3), the inaccuracy will be measured by

$$H(q,p) = \ln \sigma\sqrt{2\pi}\,[1 - \Phi(z_0)] + \frac{1}{2}\left[\frac{\delta}{\sigma} + \frac{\varphi(z_0)}{1 - \Phi(z_0)}\right]^2 \qquad (3.4.4)$$

The second term will equal zero when the true content lies at the value of the parameter μ. Moreover $H(q,p)$ will decrease with decreasing values of σ. Then, of course, either the truncation is negligible or the true content lies close to the determination limit.

However, no similar formula for $H(q,p)$ is obtainable, in order to appreciate the log-normal distribution of the results in the described way.

3.5 ANALYTICAL QUALITY CONTROL

Analytical quality control is one of most important application areas for analytical chemistry. Here the situation is such that the content of the component to be determined is supposed to lie within tolerance limits but we do not exclude in advance the probability that it may lie outside these limits. Typically, analytical quality control is carried out to find out whether the content of the analyte lies within the prescribed limits or does not lie within them. In our monograph (1979) we showed a suitable choice of the a priori distribution and presented the value of the divergence measure under normal distribution of the results.

Here we wish to mention a sequential sampling technique based on information theory as an alternative to conventional quality control techniques. It is continuous in the sense that it is performed continuously until a decision is made, when applied as a single sampling plan or as an on-line device for production lines.

First the process under consideration can be under control (state H_1) or out of control (state H_2). Being in either state, the process may yield products with contents of the controlled component outside tolerance limits with probability P_i (i=1,2) and products with acceptable contents with probability $1-P_i$. Measurements are taken sequentially and cumulative outcomes n_1 of type 1 and n_2 of type 2 are substituted into "evidence" equations corresponding to each of the states. These equations appear as

$$Ev(\frac{H_i}{n}) = Ev(H_i) + n_1 \log \frac{P_1}{P_2} + n_2 \log \frac{1-P_1}{1-P_2}, \quad i=1,2 \quad (3.5.1)$$

(Baram, 1975; Tribus, 1969) where $Ev(H_i/n)$ and $Ev(H_i)$ are evidences (negative information of odds of H_i/n or of H_i). Thus, e.g.,

$$Ev(H_1) = \log \frac{P(H_1)}{P(\overline{H}_1)} = \log \frac{P(H_1)}{P(H_2)}$$

Whenever the calculated evidence $Ev(H_1/n)$ exceeds a predetermined risk level, hypothesis of the process being in state H_1 is accepted - the process is under control,with P_1 or less probability of producing contents outside the limits. Whenever the evidence value falls below a minimum predetermined risk, level H_1 is rejected – the process is out of control, and the fraction of inconvenient outcomes has moved to at least P_2, so that some correction in the production is to be made. As long as the evidence equation yields values between the upper and lower limits, the testing is continued.

The choice of units in (3.5.1) is quite arbitrary. Log 10 tables are most widely available and then the values are expressed in decit units. Risk levels of $\pm$ 2 decits represent odds of 100 to 1 in favour of making the right decision (log 100 = 2). This level may be considered as equivalent to the classical 1% level of significance. In order to determine the initial probability that the process is in state H_i (needed for equation (3.5.1)) we can use, for instance, past records or some kind of experience. If, e.g., the prior probability that the process is in state H_1 is 0.8, the odds are 4 to 1, in starting the production, that the process is under control. Although the selection of the initial probability may be sometimes artificial, the effect of this value becomes less important as the sampling procedure progresses and more information is being accumulated.

Since the evidence of being in one state equals the negative evidence of being in the other state, only one equation from (3.5.1) is used. The decision to accept one hypothesis implies the rejection of the second one. The results of the test are plotted in a chart with $n = n_1 + n_2$ inspected contents on the horizontal axis and with evidence values (in decits) as ordinates.

If compared to the classical sequential sampling technique (Wald, 1952), in either case the acceptance, rejection and no decision regions are separated by straight lines. However, whereas in the classical technique the boundary lines are angular the approach just outlined yields horizontal lines. In the use of both techniques the risk level and the acceptable and rejectable quality levels P_1 and P_2 must be selected to determine the equations of the upper and lower limit levels.

Information theory has brought not only new methods of judging and optimizing the process of quality control (Baram, 1975; Danzer and Eckschlager, 1978; Eckschlager, 1977 and 1978; Eckschlager and Štěpánek, 1979) but it has shown another approach and fundamental philosophy. In spite of the relevance of information, provided by particular examination methods, being for the reliability of decision making on the quality and consuming value of products at least as important as its content, it is so far not taken into account in quantitative evaluation of testing processes. A suggestion how to proceed in case when information about several quality features cannot be taken for equally relevant or when relevance of particular information depends on the precision, with which the information is obtainable, can be seen in the use of the appurtenance function based on the idea of a fuzzy-subset of relevant information (compare Section 4.3).

REFERENCES

Baram, G. (1975). Sequential sampling plans based on information theory. J.Qual.Techn. 7, 20-27.

Büttner, J. (1982). Grundlagen der Anwendung der Informationstheorie auf qualitative klinisch-chemische Untersuchungen. J.Clin.Chem.Clin.Biochem. 20, 477-490.

Cleij, P., and Dijkstra, A. (1979). Information theory applied to qualitative analysis. Fresenius Z.Anal.Chem. 298, 97-109.

Danzer, K. (1974). Informationstheoretische Charakterisierung von Verteilungsanalysen. Z.Chem. 14, 73-75.

Danzer, K., and Eckschlager, K. (1978). Information effieciency of analytical methods. Talanta 25, 725-726.

Doerffel, K., and Eckschlager, K. (1981). Optimale Strategien in der Analytik. Verlag Harry Deutsch, Thun, Frankfurt/M.

Eckschlager, K., and Vajda, J. (1974). Amount of information of repeated higher precision analyses. Coll.Czechoslov.Chem.Commun. 39, 3076-3081.

Eckschlager, K. (1975). Informationsgehalt analytischer Ergebnisse. Fresenius Z.Anal.Chem. 277, 1-8.

Eckschlager, K. (1976). Information content of instrumental analysis results. Coll.Czechoslov.Chem.Commun. 41, 1875-1878.

Eckschlager, K. (1977). Information theory as applied to chemical analysis. Anal.Chem. 49, 1265-1267.

Eckschlager, K. (1978). Information content of analytical quality control. Coll.Czechoslov.Chem.Commun. 43, 231-238.

Eckschlager, K., and Štěpánek, V. (1978). Information content of trace analysis results. Mikrochim. Acta 1978, I., 107-114.

Eckschlager, K. (1979). Information content of analytical results subject to systematic error. Coll.Czechoslov.Chem.Commun. 44, 2373-2377.

Eckschlager, K., and Štěpánek, V. (1979). Information theory as applied to chemical analysis. J.Wiley-Interscience, New York.

Eckschlager, K., and Štěpánek, V. (1980). Accuracy of analytical results. Coll.Czechoslov.Chem.Commun. 45, 2516-2523.

Eckschlager, K. (1981). Information content of analytical signals of instrumental methods. Coll.Czechoslov.Chem.Commun. 46, 478-483.

Eckschlager, K., and Štěpánek, V. (1981). Information theory approach to trace analysis. Mikrochim. Acta 1981, II., 143-150.

Eckschlager, K. (1982). Information properties of an analytical system. Coll.Czechoslov.Chem.Commun. 47, 1580-1587.

Eckschlager, K., and Štěpánek, V. (1982). Properties of the divergence measure of information content as related to quantitative analyses. Coll.Czechoslov.Chem.Commun. 47, 1195-1202.

Eckschlager, K., and Štěpánek, V. (1982). Information theory in analytical chemistry. Anal.Chem. 54, 1115A-1127A.

Frank, J., Veress, G., and Pungor, E. (1982). Some problems of the application of information theory in analytical chemistry. Hungar.Sci.Instr. 54, 1-9.

Kaiser, H. (1970). Quantitation in analytical chemistry. Anal.Chem. 42 (2), 24A; (4) 26A.

Kaiser, H. (1972). Zur Definition von Selektivität, Spezifität und Empfindlichkeit von Analysenmethoden. Fresenius Z.Anal.Chem. 260, 252-256.

Liteanu, C., and Rica, J. (1979). Utilization of the amount of information in evaluation of analytical methods. Anal.Chem. 51, 1986-1995.

Štěpánek, V., and Eckschlager, K. (1979). Information content of chemical analysis results and methods. in The proceedings of the Sixth International Codata Conference, Pergamon Press, Oxford and New York.

Tribus, M. (1969). Rational descriptions decisions and design. Pergamon Press, Elmsford.

Vajda, J., and Eckschlager, K. (1980). Analysis of a measurement information. Kybernetika 16, 120-144.

Wald, A. (1952). Sequential analysis. J.Wiley, New York.

Zadeh, L.A. (1965). Fuzzy sets. Information and Control 8, 338-353.

Zadeh, L.A. (1968). Fuzzy algorithms. Information and Control 12, 94-102.

CHAPTER 4
Investigation of Analytical Systems

4.1 SOURCES OF UNCERTAINTY

Until now our considerations have been always concerned with information-theoretical qualification of either a specific analytical method (or a combination of several methods) or of specific types of analyses (e.g., of identification, quantitative analyses, etc.). A problem which depends upon knowledge of chemical composition is always solved in an analytical system the subsystem of which, not standing alone, is an analytical method. Attention to the usefulness of a systems approach to solving analytical problems was drawn in 1972 by Malissa. An excellent and lucid description of an analytical system, serving not only to obtaining information about an analysed object but also quality assessment of obtained information, was presented by Taylor (1981); his paper is also the grounds of another contribution to the idea of an analytical system (Borman, 1982). In Figure 7 a simplified scheme of an analytical system is illustrated; a specific problem enters this system in its input. After finding out whether we really deal with a problem connected with need of information about the chemical composition, we formulate a model of the problem and a plan of its experimental solution. This plan should

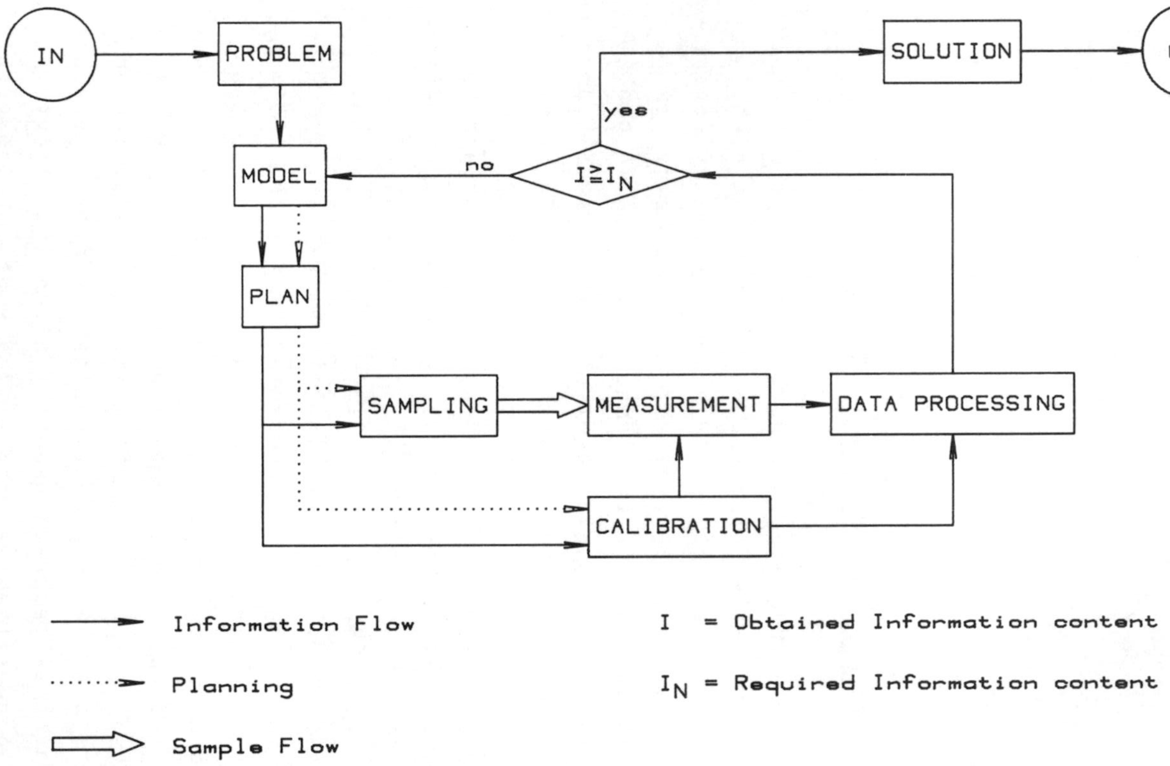

FIG. 7. Scheme of an analytical system.

contain: (i) a sampling plan; (ii) choice of analytical method and of device and a plan of the procedure of the analysis; (iii) choice of a manner of calibration and a plan of performing it (choice of reference material, a plan of calibration frequencies or of adjusting the device) and finally (iv) a plan of the way of processing measured data. After the realisation of the plan it is necessary to decide whether obtained results will be sufficient for the solution of the problem or whether it will be needed to work out a new more perfect model by the use of obtained results.

Consider now what subsystems of an analytical system have effect upon the comprehensiveness of obtained information or upon the a priori and the a posteriori uncertainties and upon the information content. The a priori uncertainty is basically given by the problem and can be formulated in the phase of forming the model. The a posteriori uncertainty will be affected by the adequacy of the sampling, of proper measurements (including the occurence of noise), of the calibration and, to some extent, also by the means of processing measured data.

We have paid attention to the sampling process in Chapter 3 in connection with our considerations about analytical quality control. Yet quite generally the sampling strategy must conform to what information we want to obtain or what uncertainty we wish to reduce. The objective of sampling is such a selection that brings a representative sample. Of course, a representative sample will be collected when we want to obtain information about average composition of the analysed material, when we seek for the composition of individual parts of a mixture and again when we want to find out, for instance, the gradient of composition in raw materials, etc.

Except for sampling of a perfectly homogeneous material the withdrawal of a sample represents a source of a posteriori uncertainty and reduces-sometimes substantially-the comprehensiveness of information about the analysed material which comes from an analytical system. If the distribution of the content of a determination of component i in material is normal around the mean value X_i with variance σ_i^2, and if the sampling distribution is also normal with mean value μ_s and with variance σ_s^2, we can qualify the inaccuracy of the representation in terms of the Kerridge-Bongard measure as

$$H(q,p) = \frac{1}{2}\left[\ln 2\pi\sigma_s^2 + \frac{(X_i - \mu_s)^2 + \sigma_i^2}{\sigma_s^2}\right] \qquad (4.1.1)$$

Obviously this inaccuracy is, for given σ_i^2, greater the larger is the ratio σ_i / σ_s and the difference $|X_i - \mu_s|$, especially for small values of σ_s (therefore it is greater, the poorer is the "representativeness" of the sample). For $X_i = \mu_s$ and $\sigma_s = \sigma_i$, i.e., for perfect "representativeness", the measure of inaccuracy in (4.1.1) depends on the heterogeneity of the material characterized by the value of σ_i. This inaccuracy can also serve in choosing or optimizing the sampling procedure.

We have shown earlier (1979) the influence of calibration upon the value of the standard deviation of measurements. The processing of data cannot create by itself new information that would not be encoded in the results of the experiment; yet it can, to the contrary, if it is carried out in an inappropriate way, enhance the a posteriori uncertainty of the results.

We work in the framework of an analytical system in order to obtain information serving to the solution of a given problem or information to be employed as basis for decision making (e.g., technical or economic decision, medical diagnosis, etc.). The total information leaving an analytical system need not be relevant for a given task. Thus we will rate an analytical system according to the amount of information that can be utilized

$$A = \sum_{i=1}^{n} I_i(q||p).k_i \qquad (4.1.2)$$

where $I_i(q||p)$ is the amount of information about identity i or about the content of the ith component (i=1,2, ..., n) and k_i is the relevance coefficient (cf. Section 3.3). We can understand this coefficient as a value of the membership function of the ith information to the fuzzy subset of relevant information. Because of $0 \leqq k_i \leqq 1$, the inequalities $0 \leqq A \leqq M(q||p)$ are valid ($M(q||p)$ has been introduced in Section 3.3).

4.2 DESCRIPTION OF ANALYTICAL CHEMICAL SYSTEMS

The introduction of information theory into analytical chemistry is closely related to the analytical system in which a particular problem (an analysis and its evaluation) is to be solved. Only in this way can we avoid misunderstandings and contradictions. Thus we have to take into account requirements necessary for treating an analytical system as a communication one, adequate models for describing such a system including probability distribution of the signals, and uncertainty introduced in interpreting the results of the analysis.

The term "communication system" refers to a system with the following structure: source-encoder-channel-decoder-result. Noise can interfere with the communication channel. When the signals emitted by the source are treated as a discrete random variable, the values of which in a sequence are independent of each other, the system is said to be without memory. In the preceding expositions we dealt with analytical systems of this kind. Besides entropies related to the material H(P(X)) and to the signal H(P(Z)) we also employed conditional uncertainty E = H(P(X|Z)) in identification and qualitative analysis. In addition conditional entropy H(P(Z|X)) and joint entropy

$$H(P(X,Z)) = \begin{cases} -\sum_i \sum_j P(X_i,Z_j) \text{ ld } P(X_i,Z_j) \\ \\ -\int_{-\infty}^{\infty} \int_{-\infty}^{\infty} f(x,z) \text{ ld } f(x,z)\, dx\, dz \end{cases} \quad (4.2.1)$$

can be defined; the latter is understood as the uncertainty of the communication system. According to (2.2.4) E = H(P(X|Z)) can be written as

$$E = H(P(X|Z)) = -\sum_i \sum_j P(Z_j)\, P(X_i|Z_j) \text{ ld } P(X_i|Z_j)$$

$$= -\sum_i \sum_j P(X_i,Z_j) \text{ ld } P(X_i|Z_j) \quad (4.2.2)$$

which is comparable to (4.2.1) in the discrete case. It is also the form in which we derived the conditional uncertainty in Section 4.3 in our monograph (1979). It can be easily seen that equalities

$$H(P(X,Z)) = H(P(X|Z)) + H(P(Z)) = H(P(Z|X)) + H(P(X))$$

hold (cf. Appendix).

Another measure related to a communication system is the average mutual information $T(X,Z)$, called trans-information of the system, which we have mentioned in Chapter 2 and which can be found in the Appendix along with its properties. The following formulae express the relationships between this measure and the various types of entropy:

$$T(X,Z) = H(P(X)) - H(P(X|Z)) \qquad (4.2.3)$$

$$T(X,Z) = H(P(Z)) - H(P(Z|X)) \qquad (4.2.4)$$

$$T(X,Z) = H(P(X)) + H(P(Z)) - H(P(X,Z)) \qquad (4.2.5)$$

These relationships are equally applicable to characterizing an analysis in terms of transinformation, depending on the available knowledge of the probability distribution of the components or of the contents and of the signals (and the noise). In practice the joint distribution is not known. Thus only relationships (4.2.3) and (4.2.4) are applicable. We have presented a special use of (4.2.3) in (2.2.7) where we evaluated the amount of information in identification and qualitative analyses as decrease of uncertainty; we provided the probabilities $P(X_i|Z_j)$ by employing the Bayes' rule to a priori known probabilities $P(Z_j|X_i)$. The relationship (4.2.4) is generally applicable, i.e., the transinformation can be obtained as the difference of the pre-measurement entropy referring to the result of measurement and of the a posteriori entropy (cf. Frank et al., 1982).

Sometimes we can find such interpretation of the information content in analytical chemistry, according to which the amount of information provided by an analysis is calculated as the difference of entropies before and after analysis respectively, in which the former is referred to the material. Thus the particular difference $H(P(X)) - H(P(Z|X))$ is introduced; this is, of course, only a matter of convention, yet it has no mathematical basis and its use may lead to misunderstandings.

In qualitative analysis, the communication system appears in the interpretation of signals with inputs being possible components. If we assume probability distributions for the inputs and/or the outputs the joint probability distribution of the communication system can be determined from the matrix of conditional probabilities and the marginal distributions. Thus the transinformation can be obtained by definition

$$T(X,Z) = \sum_i \sum_j P(X_i,Z_j) \; \mathrm{ld} \, \frac{P(X_i,Z_j)}{P(X_i) \cdot P(Z_j)}$$

The efficiency of this measure of the amount of information can be seen from the following: if the inputs and the outputs are independent, then $T(X,Z) = 0$ (because the joint distribution is the product of marginal distributions). On the other hand the maximum amount of information is provided if only diagonal elements of the matrix differ from zero, i.e., if unambiguous mapping of the inputs into the outputs exists.

In quantitative analysis an analytical system is commonly described as a communication system with additive noise. However, the noise is related to both the results and the signals. If the additivity exists in either case we can write

$$X = Z + \varepsilon$$

i.e., the input is the sum of the output and of the error, where the error is the sum of errors of measurement and of interpretation. Thus, in analytical systems regarded as communication systems with this additivity, the entropy $H(P(X|Z))$ becomes

$$H(P(X|Z)) = H(P(Z + \varepsilon|Z)) = H(P(\varepsilon))$$

and the relationship (4.2.3) for the transinformation will yield

$$T(X,Z) = H(P(X)) - H(P(\varepsilon))$$

Hence it is shown that the information content measured as decrease of uncertainties in transition from a predistribution to the distribution of the errors of analytical results (frequent in analytical literature) is a special form of the transinformation.

If we regard an analytical system as a communication one we have, after we have postulated a particular model, to set up hypotheses concerning the probability distributions of the random variables in the model. We assume a uniform a priori distribution in a finite interval unless some preliminary knowledge is available, i.e., a distribution resulting in the maximum entropy. On the other hand the errors of measurement follow a Gaussian (normal) distribution. This is an approach that we have mostly employed not only throughout preceding chapters but also in our monograph (1979) when we were dealing with another information characteristic - the divergence measure.

4.3 THE ROLE OF UNCERTAINTIES IN ANALYTICAL SYSTEMS

The uncertainty before analysis in the input of an analytical system can be understood as a feature of our assumptions reflecting any preliminary experimentally obtained knowledge of the composition of the sample to be analysed, or it can be caused by known parameters of a device (ranges of signal position and intensity). The a posteriori uncertainty depends only on the behaviour of analytical results (or signals) and is of basic importance for further reasoning. It can be calculated from the distribution of results or of signal intensities in a specific position or from the joint distribution of signal position and intensity or from conditional probabilities of the presence of individual identities given a signal in a specific position in the output or by employing partial sensitivities.

We have shown that the uncertainty after a multicomponent qualitative or quantitative analysis depends on the selectivity (cf. (3.3.2) through (3.3.4) and (3.3.1)) when we substitute conditional probabilities or partial sensitivities set together in matrices representing the input-output relation. The a posteriori uncertainty of quantitative results is explicitly affected by the characteristics of the random (σ^2) and the systematic (δ) error components and it depends also for a given precision of signal intensity measurements σ_y on the sensitivity S_i (cf. Section 3.3). In trace analyses it depends on the determination limit x_o. Some other details on the dependence of the a posteriori uncertainty on the behaviour of the results have been presented (Eckschlager, 1982).

Let us introduce the idea of an instrumental analytical method that makes possible selective identification and determination of all components present in contents from 0 % through 100 % with precision expressed by a negligibly small variance of unbiased results, i.e., the mean values of their distributions coincide with true contents of individual components. We will call this method ideal and the device enabling to carry out such analyses will be an ideal device. The results of an ideal procedure in qualitative analysis (the discrete case) have zero a posteriori uncertainty. For continuous distributions it can even be negative.

All real methods and devices show greater or smaller deviations from the ideal state as far as selectivity, precision, accuracy, sensitivity, and the determination and detection limits are concerned. All these features are connected with a posteriori uncertainty, which is the greater the more the real method differs from the ideal one. We can therefore take this uncertainty for a measure of the deviation of a real method or device from the ideal one and we denote as optimum, for a given analytical problem, that procedure or device that shows the least a posteriori uncertainty of all those taken into account. Then we understand the optimization as a process of approaching the ideal case.

The uncertainty of analytical results arises as a consequence of imperfection of chemical or physical interaction, which is the basis of the analytical procedure, or it can follow from the imperfection of the device. That enables us to proceed from a pure black-box understanding of an analytical process to the investigation of the influence of imperfection of the chemical and physical substances of this process upon its information properties. Or, in another words, to approach judgments on special properties of analytical methods and devices with the use of characteristics derived from information-theoretical grounds. However, this reasoning does not fall into the field of chemometrics and thus we do not deal with it here.

4.4 QUANTIFICATION OF FEATURES OF AN ANALYTICAL SYSTEM

Some features of analytical processes that have been so far understood only qualitatively, and other ones that were earlier described quantitatively, can be brought into relation with uncertainty after analysis and expressed on the information-theoretical basis. Thus, e.g., the precision and the accuracy of quantitative analysis results are well enough characterized as the random and the systematic error component respectively, by employing statistical quantities such as the variance and the bias. Efforts to evaluate both error components of analytical results in a single quantity led to proposals of various definitions of the, so called, total error (McFarren et al., 1970; Eckschlager, 1972; Midgley, 1977). These definitions were built on pragmatic concepts of such a measure rather than derived exactly. We have recently replaced these attempts, which did not prove acceptable, by a measure of accuracy (cf. (3.2.3) and (3.2.5)). In all cases this quantity measures the precision and the bias of the results simultaneously, and is the smaller the greater is either error component (Eckschlager and Štěpánek, 1980). In this definition of "a measure of accuracy" we can see the first example of quantifying basic features of an analytical system in an information-theoretical approach. This measure has found use also as a response function in optimizing analytical procedures.

Another feature of analytical systems classified earlier is specificity and selectivity. Kaiser (1972) defined - if we use his original notation - a measure of the selectivity of a multicomponent quantitative analysis as

$$\Xi = \min_{i=1,2,\ldots,n} \left(\frac{\gamma_{ii}}{\sum_{j=1}^{m} \gamma_{ij} - \gamma_{ii}} - 1 \right) \qquad (4.4.1)$$

where γ_{ij} is the partial sensitivity ($\gamma_{ij} = S_{ij}$) of the determination of the ith component by means of a signal in position Z_j. In a like manner Kaiser defined a measure of specificity of an analyte as

$$\Psi_a = \frac{\gamma_{aa}}{\sum_{i=1}^{n} \gamma_{ii} - \gamma_{aa}} - 1 \qquad (4.4.2)$$

The quantity Ξ can be used to express the selectivity of a procedure of qualitative or identification analysis if we substitute $\gamma_{ij} = P(X_i|Z_j)$. It is true in either case that the selectivity of an analysis is the more perfect the greater is Ξ ; indeed, if $\gamma_{ij} = P(X_i|Z_j)$, it does not assume such high values even when the selectivity is good as it does for $\gamma_{ij} = S_{ij}$ where S_{ij} can be large enough ($\approx 10^3$). The a posteriori uncertainty of a qualitative or identification analysis, i.e., the uncertainty after analysis when a signal in position Z_j in the output was measured, is given by Shannon's entropy (3.3.2) and the a posteriori uncertainty of a multicomponent quantitative analysis is given by the same formula if we substitute $h_{ij} = d_{ij}$ according to

(3.3.1), i.e., the relative partial sensitivity. As we have shown in Section 3.3 the relative entropies H_r in (3.3.4) or $H_{r(ij)}$ in (3.3.3) are more appropriate measures for selectivity or specificity. These quantities are small for good selectivity and take on zero values for perfect selectivity. The entropy can assume different values for the same value of quantity Ξ according to the spread of the γ_{ij}'s around γ_{ii}.

The selectivity and the specificity are features of an analytical system quantified for identification and qualitative analyses in terms of conditional probabilities and for quantitative analyses by the use of partial sensitivities. Nevertheless they are of the same meaning in qualitative and quantitative analyses: they represent measures of capability of the analytical system to distinguish and at the same time to prove or to determine several components different qualitatively. Therefore they are basically important in judging methods of multicomponent analysis. By the way, the number of distinguishable components is the basis of determining "the informing power" of an analytical method and of the employment of the Brillouin measure (see Section 2.1).

The basic parameters in trace analyses are the determination and detection limits, i.e., the least amount of the analyte than can be determined or qualitatively proved. Several definitions of this parameter have been published (Liteanu and Rica, 1979; Kaiser, 1966; Curry, 1968; Long and Winefordner, 1983). We have seen in Section 3.4 how the information content depends on the determination limit x_0.

Liteanu and Rica (1979) have dealt in detail with problems of detecting analytical signals; they have presented information-theoretical definitions of a decision threshold and of a detection limit. According to the latter a detection limit is the minimum value of the amount of analyte for which a single detection experiment can remove all the uncertainty existing prior to the analysis. Thus it corresponds basically to the value x_1 introduced in Section 3.1. However, this definition cannot be applied to the determination limit, because quantitative analysis always removes only some part of the uncertainty existing prior to the experiment.

Thus the determination limit x_o appears in the formulae for the information gain in trace analyses. Obviously this gain is greater when the true content is above the determination limit - and often substantially greater - than that for the case $X < x_o$. Therefore the determination limit x_o of an adopted method decides on how frequently the more favourable case $X > x_o$ will occur.

4.5 THE EFFECT OF CALIBRATION

So far our considerations about information contents of the results of analyses have been concerned only with analytical methods or laboratory devices. This reasoning is comprehensible for the evaluation, choice, and optimization of analytical methods, devices, and procedures (Doerffel and Eckschlager, 1981), yet it does not do justice to the real information properties of an analytical system as defined by Taylor (1981) and illustrated in Figure 7. The sole source of information, i.e., of the decrease of uncertainty, in this system is the measurement; of course, inevitably sampling and calibration contribute to uncertainty as well and thus decrease the comprehensiveness of information produced by the analytical system.

The divergence measure of the amount of information gained from measurement is for an a priori uniform distribution $U(x_1, x_2)$ and for an a posteriori normal distribution $N(\mu_A, \sigma_A^2)$ given according to (2.2.6) by

$$I(q||p) = \ln \frac{x_2 - x_1}{\sigma_A \sqrt{2\pi e}}$$

The standard deviation of the analysis results expressed as mass ratios or concentrations depends on the way of calibration. If the dispersion of the measurements of signal intensities is σ_y then, in the case of carrying out calibration by the method of standard addition, we obtain for the standard deviation of the results

$$\sigma_A \approx \frac{\sigma_y x_s}{(y_2 - y_1)^2} \sqrt{y_1^2 + y_2^2} \; ; \; y_2 > y_1 \qquad (4.5.1)$$

where x_s is the amount of the analyte added and y_1 and y_2 are signal intensities before and after addition respectively. This approximate formula has been derived and used in our monograph of 1979. However, it does not enable a deeper insight in order to assess circumstances under which σ_A does not increase and thus the information content does not decrease. Therefore another approach to obtaining this statistic will be shown below in the framework of the treatment of calibration straight lines.

If the calibration is performed in terms of a calibration straight line the following formula for the standard deviation of the results holds:

$$\sigma_A = \frac{\sigma_y}{b_c} \sqrt{\frac{1}{n_A} + \frac{1}{n} + \frac{\left[\bar{y}(n_A) - \bar{y}\right]^2}{b_c^2 \sum_i (x_i - \bar{x})^2}} \qquad (4.5.2)$$

Here b_c is the slope of the calibration line, $\bar{y}(n_A)$ is the mean of parallel determinations and n_A is their number, N is the number of points of which the calibration curve is constructed, and x_i is the concentration of the analyte in the ith standard (i=1,2, ..., N).

Details on both ways of calibration can be found in Section 6.5 of our monograph (1979) and in Appendix B.5 at the same place.

The standard deviation σ_A above is inversely proportional to the sensitivity b_c. The variabilities of both parameters of the calibration straight line (the intercept a_c and the slope b_c) become evident in terms σ_y^2/N and

$$[\sigma_y^2(\bar{y}(n_A) - \bar{y})^2] / \sum_i (x_i - \bar{x})^2 = \sigma_b^2(\bar{y}(n_A) - \bar{y})^2$$

entering the variance σ_A^2 while the term σ_y^2/n_A is due to the dispersion caused by the analysis. The aim is to achieve such experimental conditions that the sum under the square root in the formula for σ_A does not exceed the unit. This is primarily feasible through $n_A = 2$ and $N = 10$. Moreover the standard deviation σ_A depends on the mean $\bar{y}_A$ of the measurements although we keep σ_y for constant. This effect is the stronger, the greater is σ_b/b_c opposite to $\sigma_y/\bar{y}_A$. It can be expected that σ_y assumes its minimum in the middle of the interval $\langle x_{min}, x_{max} \rangle$.

Thus we can summarize that optimum measurement conditions are achieved (provided that calibration standards are distributed equidistantly) if the standard deviation of signal intensities is small, the sensitivity b_c is large, the calibration line is constructed from a sufficient number of points, and at least two parallel determinations are performed in the analysis. Then we avoid an undesirable increase in σ_A and we do not lower the information content of the result of such a quantitative analysis.

The standard-addition method of calibration can be viewed as a special case of the regression line $E[\eta_i] = bX_i$ passing through the origin. Here the standard deviation of the analysis result given above in (4.5.2) reduces to

$$\sigma_A = \frac{\sigma_y}{b}\sqrt{\frac{1}{n_A} + \frac{\bar{y}^2(n_A)}{b^2 \sum_i x_i^2}} \qquad (4.5.3)$$

Almost no increase of random dispersion occurs already with $n_A = 1$ provided that the second term under the square root is sufficiently small. If we calibrate with one addition x_s then the pair (x_s; $\Delta y = y_2 - y_1$) lies on the regression line and we obtain (with $n_A = 1$)

$$\sigma_A = \frac{\sigma_{\Delta y}}{b}\sqrt{1 + \frac{y_1^2}{b^2 x_s^2}} \qquad (4.5.4)$$

Because we keep σ_y constant we can substitute $\sigma_{\Delta y} = \sigma_y\sqrt{2}$ and when the calibration addition is expressed as a multiple of the content of the analyte, i.e., $x_s = qx_A$, this formula turns out to be

$$\sigma_A = \frac{\sigma_y\sqrt{2}}{b}\,\frac{\sqrt{q^2 + 1}}{q} \qquad (4.5.5)$$

This is true as far as we take both signals (before and after addition) for uncorrelated. However, some rate of correlation is closer to the reality and if we take it into account the variance of the difference $\Delta y = y_2 - y_1$ is equal to $2\sigma_y^2 - 2\,\mathrm{cov}(y_1, y_2) = 2\sigma_y^2(1 - \rho)$ where ρ is the correlation coefficient. Hence we obtain

$$\sigma_A = \frac{\sigma_y}{b} \frac{\sqrt{q^2 + 1}}{q} \sqrt{2(1 - \varrho)} \qquad (4.5.6)$$

Thus apparently a large addition and high correlation rate diminish the dispersion of the determination of the analyte. In order to achieve no increase of the relative standard deviation through calibrating by the standard-addition method when we set $x_s = x_A$ (i.e., $q = 1$)

$$\frac{\sigma_A}{x_A} = \frac{\sigma_y}{y_1} 2 \sqrt{1 - \varrho}$$

it is necessary that ϱ be at least 0.75. However, such a high correlation between y_1 and y_2 cannot be always expected; thus the standard-addition method usually provides results with rather large dispersion and therefore lowers their information content.

Remark:

Formula in (4.5.4) cannot be obtained by mathematically manipulating the formula (4.5.1) quoted above and transferred from our monograph (1979). The latter had been derived as an approximate one from the Taylor expansion of x_A as a function of two signals (the addition x_s can be assumed as a quantity missing error) without introducing a regression line while the former was obtained by simplifying a more complicated approximate formula for σ_A in general linear regression (cf. Appendix B.5 of the cited monograph).

If we treat the function $x_A = x_s y_1/(y_2 - y_1)$ similarly as in the latter case except that we take the correlation between the two measurements into consideration we apply the general approximation

$$\sigma_A^2 = \left(\frac{\partial x_A}{\partial y_1}\right)^2 \sigma_y^2 + \left(\frac{\partial x_A}{\partial y_2}\right)^2 \sigma_y^2 + 2\rho\sigma_y^2 \left(\frac{\partial x_A}{\partial y_1}\right)\left(\frac{\partial x_A}{\partial y_2}\right)$$

which yields

$$\sigma_A = \frac{\sigma_y x_s}{(y_2 - y_1)^2} \sqrt{y_1^2 + y_2^2 - 2\rho y_1 y_2}$$

By substituting the slope b and $x_s = qx_A$ we obtain

$$\sigma_A = \frac{\sigma_y}{b} \frac{q + 1}{q} \sqrt{2(1 - \rho)} \qquad (4.5.7)$$

which differs from (4.5.6). According to this formula σ_A is always somewhat greater for any q than the corresponding result calculated by (4.5.6).

Furthermore we can compare the standard-addition method with the evaluation by a calibration curve in order to recognize the correlation rate with which the standard-addition method has still sense. If follows from (4.5.2) that the error of the determination is different for different concentrations and is minimum when $\bar{y}(n_A) = \bar{y}$. In that case the error in finding the slope of the calibration line has no effect on the results of the analysis. Thus in the middle of the calibration line we have (for $n_A = 1$) $\sigma_A = (\sigma_y/b_c)\sqrt{1 + 1/N}$ and by comparing with (4.5.6) we obtain

$$\rho = 1 - \frac{N + 1}{2N} \frac{q^2}{q^2 + 1} \qquad (4.5.8)$$

The quantities in (4.5.8) are the least correlation coefficients in the presence of which the standard-addition method, with addition overpassing the content to be determined q times, brings benefit compared to the use of a calibration line constructed of N points. For average values of q and N the minimum correlation coefficients lie slightly above 0.50 while with small additions ($1 < q < 2$) the standard-addition method is preferable only when ρ is above 0.7 (Ehrlich and Gerbatsch, 1965).

The objective of calibration is to remove the systematic error. In this respect let us consider the meaning of $I(q||p)$ by (2.2.6) and of $I(r;q,p)$ by (3.2.1). It is obvious that $I(r;q,p)$ changes into $I(q||p)$ for $\delta=0$. In fact, of course, we take δ for a systematic error only when it is statistically significant whereas, in the formula (3.2.1) any non-zero value of δ lowers $I(r;q,p)$, regardless of its statistical significance. We have to understand it in the following way: if the calibration is performed in such a way that the rise of a systematic statistically significant error does not need to be anticipated we adopt $I(q||p)$ in (2.2.6) to the evaluation of the information content. However, unless it can be guaranteed by carrying out calibration that a systematic error does not occur, we have to express the information content in terms of $I(r;q,p)$ with the notion that any deviation $\delta = |X - E[\xi]|$ decreases the information content of the result.

We have already mentioned that biased results can misinform us to that extent that the information content according to (3.2.1) is zero. If we denote the length of the interval $x_2 - x_1 = w\sigma$ and the systematic error $\delta = k\sigma$ we have

$$I(r;q,p) = \ln \frac{w\sigma}{\sigma\sqrt{2\pi e}} - \frac{1}{2}\left(\frac{k\sigma}{\sigma}\right)^2 = \ln \frac{w}{\sqrt{2\pi e}} - \frac{1}{2}k^2$$

which yields $I(r;q,p) = 0$ for $k = \sqrt{\ln(w^2/2\pi e)}$. If we substitute $k = z(\alpha)$, where $z(\alpha)$ is a critical value of the standardized normal distribution, we can determine the significance level α for which it holds that an error δ significant at this level depreciates the result of the analysis. Obviously $(1 - \alpha)$ will be the greater the larger is $w = (x_2 - x_1)/\sigma$. Several values are shown in Table 4.

TABLE 4. Significance levels of errors δ in dependence on $\underline{w}$.

w	$k = z(\alpha)$	$1 - \alpha$
6.0	0.86	0.610
8.0	1.15	0.750
12.0	1.46	0.856
20.0	1.78	0.925
50.0	2.23	0.974
100.0	2.52	0.988
372.1	3.00	0.997

With large a priori and small a posteriori uncertainties when the value of $\underline{w}$ is large, obviously a great mean error would have to arise so that the information content may be zero (see Figure 8.).

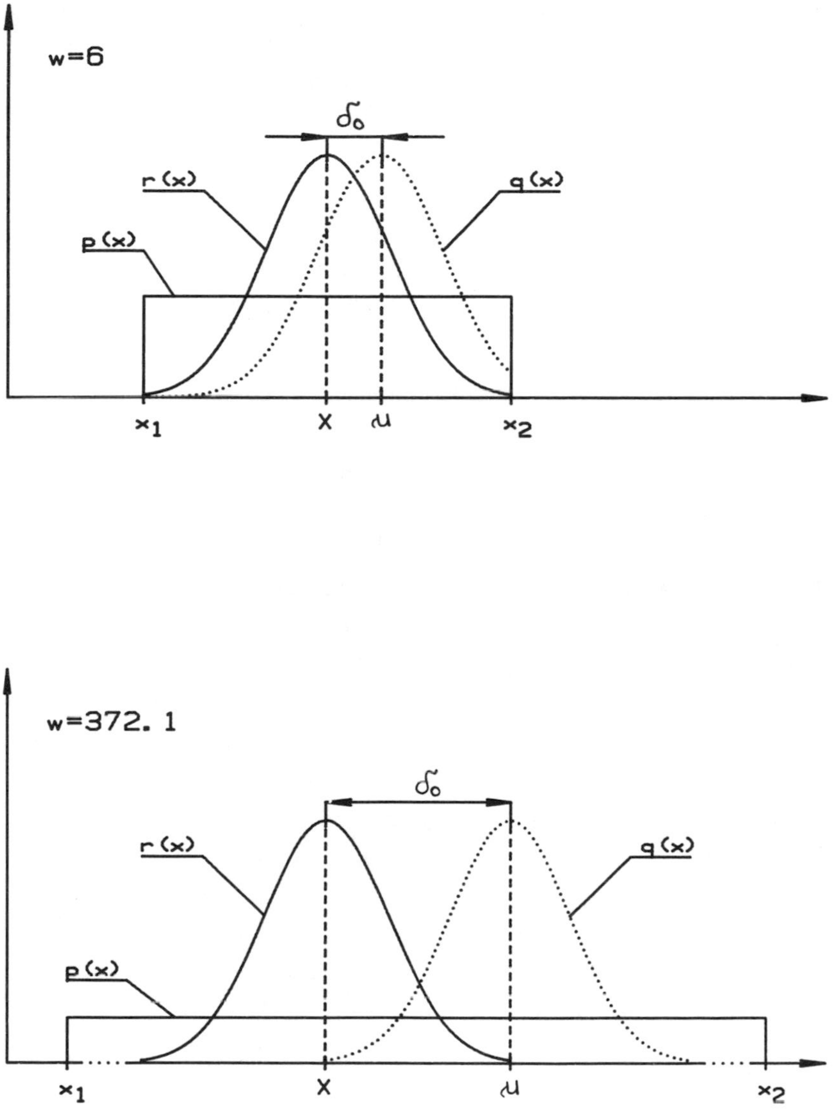

FIG. 8. δ_0 as a mean error that causes I(r;q,p)=0 for a particular ratio $w = (x_2 - x_1)/\sigma$.

4.6 SAMPLING FOR CHEMICAL ANALYSIS

In our preceding considerations about the effects upon the information content of analytical measurements we omitted that of the sampling process. A great deal of uncertainties may be hidden in a poor sampling plan. Accordingly, sampling uncertainty is often treated separately from other uncertainties in an analysis. Thus the overall variance σ^2 can be decomposed into the variance for the sampling operation σ_s^2 and that for the remaining analytical operations σ_A^2. Whenever possible, measurements should be conducted in such a way that the components of variance arising from sample variability and measurement variability can be separately evaluated. If the measurement process is under statistical control so that σ_A is already known, σ_s can be evaluated from σ found by the analysis of the samples. Otherwise, an appropriate series of replicate measurements or replicate samples can be devised to permit evaluation of both standard deviations. If the sampling uncertainty is large and cannot be reduced only a less precise analytical method is sufficient. These problems can be best demonstrated in sampling bulk materials.

Among requirements laid upon a sample by the analytical chemist is its "representativeness" which connotes a single sample from a population that can be expected to exhibit average properties of the population. However, this involves the sampling of truly homogeneous materials and the processes of homogenizing the population prior to sampling may be difficult and ordinarily unjustified. Thus a properly designed and executed random sampling plan should be preferred as it ascertains the validity of the properties of sample statistics. Or a special sampling technique is to be used when the material is segregated.

A question that is inherent in the sampling technique is the minimum number of individual samples. Unless the population is known to be homogeneous sufficient replicate samples (increments) must be analysed. To determine the minimum number of sample increments we need first to know the sampling variance σ_s^2 usually from previous experiments made on the material. The number of samples necessary to achieve a given confidence level can be estimated from

$$n = \frac{t^2 \sigma_s^2}{R^2 \mu_s^2} \qquad (4.6.1)$$

where t is a critical value of the Student distribution at the confidence level desired, μ_s is also taken from the previous knowledge of the material, and R is the percent relative standard deviation acceptable in the average. Since this expression is applicable if the sought-for component is distributed normally we can start with setting the value of t at 1.96 for 95 % confidence limits and calculate n. The t value for this preliminary n is then found in the tables, substituted, and the process is iterated to constant n.

In statistical work it is often possible to estimate the average of the population with greater accuracy than in terms of sample variance, with the same sample size and with the same total expenditure. To do this, we must replace the random sampling by choosing the sample in a special way. This is feasible in some kinds of chemical analyses, e.g., in analysing bulk materials when we are able to divide the population into a number of parts, which we conventionally call strata. It is desirable that the contents of the analyte to be measured vary as little as possible from individual to individual within the same stratum and, therefore, the mean values of different strata vary as much as possible. Here, similarly as in one-way classification in the analysis of variance, this decomposition is true:

$$\sum_{i=1}^{k} \sum_{j=1}^{N_i} (x_{ij} - \mu)^2 = \sum_{i=1}^{k} \sum_{j=1}^{N_i} (x_{ij} - \mu_i)^2 + $$

$$+ \sum_{i=1}^{k} N_i(\mu_i - \mu)^2 \qquad (4.6.2)$$

where k is the number of strata, N_i is the number of individuals (increments) in ith stratum, x_{ij} is the value of the random variable (the content of the analyte) measured on jth individual in ith stratum, μ_i is the average in ith stratum, μ is the average in the population and

$$N = \sum_{i=1}^{k} N_i$$

is the number of individuals in the population.

By random sampling within strata we can estimate each stratum mean with accuracy depending on the within strata variation. These estimators can then be used for the estimate of the population mean and its variance. If we have succeeded in choosing strata so that the variance of the estimator of μ_i is small for each i, then the estimator of the population mean will also have a small variance.

Suppose we take random samples of size $n_1, n_2, \ldots, n_k$ from the first through the *k*th stratum respectively and measure the value of the random variable ξ (e.g., the concentration) for each individual in these samples. We will denote x_{ir} the value for the *r*th individual in the sample from the *i*th stratum and *n* the total sample size so that

$$n = \sum_{i=1}^{k} n_i.$$

Then the sample mean

$$\bar{x}_i = n_i^{-1} \sum_{r=1}^{n_i} x_{ir}$$

is an unbiased estimator of the stratum mean μ_i and

$$\bar{x} = N^{-1} \sum_{i=1}^{k} N_i \bar{x}_i$$

is an unbiased estimator of the population mean μ.

According to the results known from the finite population sampling (see, e.g., Wilks, 1963)

$$\sigma^2(\bar{x}_i) = \left(\frac{1}{n_i} - \frac{1}{N_i}\right) \sigma_i^2 \qquad (4.6.3)$$

where σ_i^2 is defined as

$$\sigma_i^2 = (N_i - 1)^{-1} \sum_{j=1}^{N_i} (x_{ij} - \mu_i)^2 \qquad (4.6.4)$$

and we obtain for the variance of $\bar{x}$

$$\sigma^2(\bar{x}) = N^{-2} \sum_{i=1}^{k} N_i^2 \sigma^2(\bar{x}_i) = N^{-2} \sum_{i=1}^{k} N_i^2 \left(\frac{1}{n_i} - \frac{1}{N_i}\right) \sigma_i^2$$

$$= N^{-2} \sum_{i=1}^{k} \frac{N_i^2}{n_i} \sigma_i^2 - N^{-2} \sum_{i=1}^{k} N_i \sigma_i^2 \qquad (4.6.5)$$

If the sample sizes are <u>proportional</u> to the sizes N_i of the strata, so that $n_i = nN_i/N$, the formula in (4.6.5) is simplified to

$$\sigma^2(\bar{x}) = \left(\frac{1}{n} - \frac{1}{N}\right)\left(\frac{1}{N} \sum_{i=1}^{k} N_i \sigma_i^2\right) \qquad (4.6.6)$$

If a random sample of size $\underline{n}$ had been chosen from the whole unstratified population the variance of the mean of the sample values would be

$$\left(\frac{1}{n} - \frac{1}{N}\right) \sigma_o^2$$

where

$$\sigma_o^2 = (N - 1)^{-1} \sum_{i=1}^{k} \sum_{j=1}^{N_i} (x_{ij} - \mu)^2 \qquad (4.6.7)$$

If we insert the decomposition from (4.6.2) and employ the definition of σ_i^2 in (4.6.4) we obtain

$$\sigma_o^2 = (N-1)^{-1}\left[\sum_{i=1}^{k}(N_i-1)\sigma_i^2 + \sum_{i=1}^{k} N_i(\mu_i-\mu)^2\right] \quad (4.6.8)$$

Since it is a very frequent case that the stratum sizes are fairly large, then

$$(N - 1)^{-1}\sum_{i=1}^{k}(N_i - 1)\sigma_i^2 \doteq N^{-1}\sum_{i=1}^{k} N_i\sigma_i^2$$

and

$$\sigma_o^2 \geq N^{-1}\sum_{i=1}^{k} N_i\sigma_i^2$$

so that

$$(\frac{1}{n} - \frac{1}{N})\sigma_o^2 \geq \sigma^2(\bar{x})$$

where $\sigma^2(\bar{x})$ is given in (4.6.6). Thus we can usually expect that a proportional sample will provide an estimate of the true content μ of a compound with smaller variance than a simple random sample of the same size.

It is worthwhile to notice that a proportional sample can be constructed with no prior knowledge of the way in which the random variable ξ (the content) varies from stratum to stratum, or within strata. Thus the same proportional sample can be repeated to produce an increase in accuracy (as compared with unrestricted sampling) for any one of a series of variables.

Moreover we should note that the estimator $\bar{x}$ above becomes for a proportional sample

$$\bar{x} = n^{-1} \sum_{i=1}^{k} n_i \bar{x}_i$$

and so $\bar{x}$ is the sample arithmetic mean.

However, there exists still another way how to reduce the variance of $\bar{x}$ given in (4.6.5). Apparently the only part of this formula which depends on the sample sizes n_i is

$$\sum_{i=1}^{k} \frac{N_i^2}{n_i} \sigma_i^2.$$

Thus if the n_i's are chosen to minimize this quantity they will also minimize $\sigma^2(\bar{x})$.

Suppose that the total sample size $\underline{n}$ is given and that we wish to minimize $\sigma^2(\bar{x})$. Then we have to minimize the sum above, subject to the conditions $n_i \geq 0$,

$$\sum_{i=1}^{k} n_i = n.$$

This problem can be solved by finding an unconditioned minimum of the function

$$f(n_1, n_2, \ldots, n_{k-1}) = \sum_{i=1}^{k-1} \frac{N_i^2}{n_i} \sigma_i^2 + N_k^2 (n - \sum_{i=1}^{k-1} n_i)^{-1} \sigma_k^2$$

where we put $n_k = n - n_1 - n_2 - \ldots - n_{k-1}$.
Differentiating with respect to n_i and equating to zero, we find

$$\frac{N_i^2 \sigma_i^2}{n_i^2} = \frac{N_K^2 \sigma_k^2}{n_k^2} \qquad \text{for } i=1,2, \ldots, k-1$$

Therefore n_i are to be proportional to $N_i \sigma_i$ $(i=1,2, \ldots, k)$ and the appropriate values

$$n_i = n \frac{N_i \sigma_i}{\sum_{i=1}^{k} N_i \sigma_i} \qquad (4.6.9)$$

give optimum sampling for the variable concerned. Since formula (4.6.9) gives generally fractional values we use the nearest integer. If we substitute these sample sizes into (4.6.5) we obtain for the variance of the estimate of the population mean

$$\sigma^2(\bar{x}) = N^{-2} \left[\frac{\sum_{i=1}^{k} (N_i \sigma_i)^2}{n} - \sum_{i=1}^{k} N_i \sigma_i^2 \right] \qquad (4.6.10)$$

It should be noted that the constitution of an optimum sample depends on the values σ_i $(i=1,2, \ldots, k)$ and is therefore not the same for different random variables. As we have seen above, proportional sampling does not suffer from this drawback.

In many practical problems in chemical analysis the N_i are sufficiently large to be considered infinite. Then the proportions N_i/N in the strata approach the probabilities p_i of obtaining a value from the ith stratum. If these probabilities are known we obtain

$$\bar{x} = \sum_{i=1}^{k} p_i \bar{x}_i$$

and, provided the variances σ_i^2 are also known, it follows for the variance of this estimator from (4.6.10)

$$\sigma^2(\bar{x}) = \frac{1}{n} \left(\sum_{i=1}^{k} p_i \sigma_i \right)^2$$

since the second term in (4.6.10) disappears as N tends to infinity.

In this case it is easy to demonstrate the superiority of the stratified sampling over the random sampling from the entire population. If we consider the discrete uniform distribution $p_i = 1/k$ (i=1,2, ..., k), i.e., no stratum is preferred, and if $\sigma_1^2 = \sigma_2^2 = \ldots = \sigma_k^2 = \sigma^2$, then we have in stratified sampling $\sigma^2(\bar{x}) = \sigma^2/n$. Now we want to compare this measure of variability with the variance of the mean of a random sample from the entire population. Evidently we can write the variance of the population

$$\sigma_P^2 = \sum_{i=1}^{k} p_i E[(\xi_i - \mu)^2]$$

where E is the operator of expectation of a random variable (cf., Eckschlager and Štěpánek, 1979) and ξ_i is the random variable (e.g., the concentration) in the *i*th subpopulation. This variance can be adjusted and decomposed as

$$\sigma_P^2 = \sum_{i=1}^{k} p_i E\left[\{(\xi_i - \mu_i) + (\mu_i - \mu)\}^2\right]$$

$$= \sum_{i=1}^{k} p_i E[(\xi_i - \mu_i)^2] + \sum_{i=1}^{k} p_i(\mu_i - \mu)^2$$

$$= \sum_{i=1}^{k} p_i \sigma_i^2 + \sum_{i=1}^{k} p_i(\mu_i - \mu)^2$$

Since $p_i = 1/k$ and $\sigma_i = \sigma$, we obtain

$$\sigma_P^2 = \sigma^2 + \frac{\sum_{i=1}^{k} (\mu_i - \mu)^2}{k} \tag{4.6.11}$$

Thus the variance of the mean $\bar{x}$ of a random sample of size $\underline{n}$

$$\sigma_{\bar{x}}^2 = \frac{\sigma_P^2}{n} = \frac{\sigma^2}{n} + \frac{\sum_{i=1}^{k} (\mu_i - \mu)^2}{nk} \tag{4.6.12}$$

and hence $\sigma_{\bar{x}}^2 \geq \sigma^2(\bar{x})$ where $\sigma^2(\bar{x}) = \sigma^2/n$ (see above). The two variances are equal only when the strata means μ_i are all equal. The validity of the inequality can be proved for any other probability distribution.

It should be noted that the nature of the stratification does not play any role in the results achieved. However, the reduction of variance in stratified sampling is due to the assumption of a knowledge either of the sizes of the strata or of the probability distribution over the strata and of a knowledge of σ_i^2, which is not necessary in random sampling. Therefore we reduce the variance of the estimate of a true content in quantitative chemical analysis by adopting additional information about the population. On the other hand, if this information is incorrect the procedure may result in a biased estimate, which would not occur if we employed a completely random sample. However, sometimes the information about the variability of the content of the analyte in individual segments of a lot is more important than that about the average content.

In general, it is better to use stratified random sampling rather than unrestricted random sampling, provided the number of strata (segments) is not so large that only one or two samples can be analysed from each stratum. If we keep the number of strata sufficiently small that several samples can be taken from each, possible variation within the parent population can be detected and assessed without increasing the standard deviation of the sampling step. In order to obtain a valid sample of a stratified material it is desirable to divide major strata into real or imaginary subsections and select the required number of samples randomly (with the aid of a table of random numbers). If stratification is known to be absent, then measurement time and effort can be saved by combining all the samples and mixing thoroughly to produce a composite sample for analysis.

If the content of a material is dependent on time we need to know the function $x_A(t)$, which can be estimated from a series of discrete measurements or is known from physico-chemical laws or can be of stochastic nature. The relative time interval $A_t = (t_{end} - t_o)/t_A$ where t_A is the time necessary for carrying out one analysis, corresponds to the number of analyses to be performed in the period $\langle t_o, t_{end} \rangle$. In many cases $A_t = 3$ is considered to be the minimum. This value has effect upon the information content in determining the function x(t) so that the divergence measure in (2.2.6) is to be multiplied by A_t. Because of the requirement $A_t \geqq 3$ the time-dependent analytical results provide higher information content.

REFERENCES

Borman, A.S. (1982). Future of analytical chemistry. Anal.Chem. 54, 1354-1356.

Currie, L.A. (1968). Limits of qualitative detection and quantitative determination. Anal.Chem. 40, 586-593.

Doerffel, K., and Eckschlager, K. (1981). Optimale Strategien in der Analytik. Verlag Harry Deutsch, Thun, Frankfurt/M.

Eckschlager, K. (1972). Criterion for judging the acceptability of analytical methods. Anal.Chem. 44, 878-879.

Eckschlager, K., and Štěpánek, V. (1979). Information theory as applied to chemical analysis. J.Wiley-Interscience, New York.

Eckschlager, K., and Štěpánek, V. (1980). Accuracy of analytical results. Coll.Czechoslov.Chem.Commun. 45, 2516-2523.

Eckschlager, K. (1982). Information properties of an analytical system. Coll.Czechoslov.Chem.Commun. 47, 1580-1587.

Ehrlich, G., and Gerbatsch, R. (1965). Untersuchungen zur Anwendung des Eichzusatzverfahrens. Fresenius Z.Anal.Chem. 209, 35-46.

Frank, J., Veress, G., and Pungor, E. (1982). Some problems of the application of information theory in analytical chemistry. Hungar.Sci.Instr. 54, 1-9.

Kaiser, H. (1965). Zum Problem der Nachweisgrenze. Fresenius Z.Anal.Chem. 209, 1-18.

Kaiser, H. (1972). Zur Definition von Selektivität, Spezifität und Empfindlichkeit von Analysenmethoden. Fresenius Z.Anal.Chem. 260, 252-256.

Liteanu, C., and Rica, J. (1979). Statistical theory and methodology of trace analysis. E.Horwood, Chichester.

Long, G.L., and Winefordner, J.D. (1983). Limit of detection. Anal.Chem. 55, 712A-718A.

Malissa, H. (1972). Automation in und mit der analytischen Chemie. Verlag der Wiener Medizinischen Akademie, Wien.

McFarren, E.F., Lishka, R.J., and Parker, J.H. (1970). Criterion for judging acceptability of analytical methods. Anal.Chem. 42, 358-365.

Midgley, D. (1977). Criterion for judging the acceptability of analytical methods. Anal.Chem. 49, 510-512.

Taylor, J.K. (1981). Quality assurance of chemical measurements. Anal.Chem. 53, 1588A-1596A.

Wilks, S.S. (1963). Mathematical statistics. J.Wiley, New York.

Appendix

A.1

The entropy of a discrete random variable with probabilities $P(x_i) = p_i$ ($i=1,2, \dots, n$) is defined (in the finite case) as

$$H(P(X)) = - \sum_{i=1}^{n} p_i \operatorname{ld} p_i \qquad (A.1.1)$$

and enumerated in bits (binary information units). Obviously $H(P(X)) \geqq 0$ and it is finite. It may be regarded as a measure of uncertainty or of the amount of information pertaining to a system of events with these probabilities.

If a discrete random variable takes on the countable number of values x_i ($i=1,2,3, \dots$) the summation in (A.1.1) is from one to infinity and the sum need not be finite.

By definition, the entropy of a continuous random variable having probability density $p(x)$ is

$$H(p(x)) = - \int_{-\infty}^{\infty} p(x) \operatorname{ld} p(x)\, dx \qquad (A.1.2)$$

Here $H(p(x))$ can be also negative and infinite as well.

A.2

If two random variables are considered further types of entropies can be introduced. First, let ξ and η be two discrete random variables having positive probabilities in the points x_i $(i=1,2, \ldots, m)$ and y_j $(j=1,2, \ldots, n)$, respectively. Let us write for simplicity

$$\begin{aligned} P(x_i) &= p_i \\ P(y_j) &= q_j \\ P(x_i, y_j) &= r_{ij} \\ P(x_i | y_j) &= p_{i|j} \quad j=1,2, \ldots, n \\ P(y_j | x_i) &= q_{j|i} \quad i=1,2, \ldots, m \end{aligned}$$

According to well known relationships in probability theory, the joint probability is related to either conditional probability by the following equations

$$r_{ij} = p_i q_{j|i} = q_j p_{i|j}$$

and moreover

$$\sum_{j=1}^{n} r_{ij} = p_i \quad i=1,2, \ldots, m$$

and

$$\sum_{i=1}^{m} r_{ij} = q_j \quad j=1,2, \ldots, n$$

The conditional entropy $H(P(X|Y_j))$ is the uncertainty

contained in ξ, given that η has assumed a particular value y_j; its expectation is

$$H(P(X|Y)) = \sum_{j=1}^{n} q_j \; H(P(X|Y_j))$$

$$= - \sum_{j=1}^{n} q_j \sum_{i=1}^{m} p_{i|j} \; \mathrm{ld} \; p_{i|j}$$

$$= \sum_{j=1}^{n} \sum_{i=1}^{m} r_{ij} \; \mathrm{ld} \; \frac{q_j}{r_{ij}}$$

On the other hand if $H(P(X,Y))$ denotes the entropy of the joint distribution of ξ and η, i.e.,

$$H(P(X,Y)) = - \sum_{i=1}^{m} \sum_{j=1}^{n} r_{ij} \; \mathrm{ld} \; r_{ij}$$

we can see easily that

$$H(P(X,Y)) = H(P(Y)) + H(P(X|Y)) \tag{A.2.1}$$

because

$$\sum_{j} \sum_{i} r_{ij} \; \mathrm{ld} \; q_j = \sum_{j} q_j \; \mathrm{ld} \; q_j \sum_{i} p_{i|j} =$$

$$= \sum_{j} q_j \; \mathrm{ld} \; q_j \; . \; 1 = - H(P(Y)).$$

The interpretation of (A.2.1) is as follows:

the uncertainty contained in the pair of values (x_i, y_j) is the sum of the uncertainty contained in the value y_j and of the conditional uncertainty contained in the value of ξ provided that η has taken on a specific value.

In a like manner we may show that

$$H(P(X,Y)) = H(P(X)) + H(P(Y|X)) \qquad (A.2.2)$$

From definition it follows

$$H(P(X,Y)) = H(P(Y,X))$$

and furthermore

$$H(P(X)) \geqq H(P(X|Y))$$
$$H(P(Y)) \geqq H(P(Y|X))$$

The properties and relationships above can be extended to a pair of discrete random variables with countable number of values, and to a pair of continuous random variables in an analogous manner.

A.3

The transinformation (mutual information) is another measure that may be regarded as providing information about the value of one random variable contained in the value of the other. It is defined as

$$T(X,Y) = \sum_i \sum_j P(x_i, y_j) \ \mathrm{ld} \frac{P(x_i, y_j)}{P(x_i) . P(y_j)} \qquad (A.3.1)$$

in the discrete case and as

$$T(X,Y) = \int_{-\infty}^{\infty}\int_{-\infty}^{\infty} p_{\xi\eta}(x,y)\ \mathrm{ld}\ \frac{p_{\xi\eta}(x,y)}{p_{\xi}(x).p_{\eta}(y)}\ dx\ dy \qquad (A.3.2)$$

for a continuous joint distribution with p(x) and p(y) being marginal distributions.

The transinformation has the following properties (in notation for discrete random variables):

(i) $T(X,Y) = T(Y,X)$ (symmetry)

(ii) $T(X,X) = H(P(X))$

It means that the information expected from the random variable ξ is the information contained in its values about itself.

(iii) $T(X,Y) \leqq H(P(X))$, $T(X,Y) \leqq H(P(Y))$

Thus the information contained in the value of η about ξ is not greater than either the information contained in the value of ξ or the information contained in the value of η.

(iv) $T(X,Y) \geqq 0$ (non-negativity)

(v) $T(X,Y) = 0$ if ξ and η are independent

It follows that T(X,Y) is minimum if ξ and η are independent. $T(X,Y) = 0$ means that η does not contain any information about ξ and inversely.

The properties above certainly fulfil what we intuitively expect from information about, say, an experiment contained in another experiment. Therefore we omit exact proofs of them.

The relationships between the transinformation and various types of entropy are as follows

$$T(X,Y) = H(P(X)) - H(P(X|Y))$$
$$T(X,Y) = H(P(Y)) - H(P(Y|X))$$
$$T(X,Y) = H(P(X)) + H(P(Y)) - H(P(X,Y))$$

It follows that $T(X,Y) = T(Y,X) = H(P(Y))$ if, and only if,

$H(P(Y|X)) = 0$. It means that η contains as much information about ξ as about itself only if the conditional entropy of η with respect to ξ is zero, i.e., if the result of an experiment with probabilities $P(X_i)$ completely determines that of experiment with random variable η.

Finally it can be proved that the validity of $T(X,Y) = H(P(Y))$ and $T(X,Z) = 0$ implies $T(Y,Z) = 0$; which can be verbalized in the following way. If η contains maximum information about ξ then it does not contain information about any experiment with a random variable ζ independent of ξ, i.e., it does not contain any information irrelevant to ξ. This again coincides well with intuitive thinking.

A.4

A central problem in both mathematical statistics and information theory is to judge and compare two probability models (X,P) and (X,Q) of the same real source, in which X is the sample space and P and Q are probability distributions. The aim is to measure the similarity or dissimilarity with a single number. In information theory those numerical measures of divergence are of importance that are defined in terms of a particular convex function $\underline{f}$ mapping from $(0,\infty)$ into R (the set of real numbers). A deep-rooted name for them is <u>f-divergences</u>: $D_f(q||p)$ where $\underline{p}$ and $\underline{q}$ are functions of the probability distribution (in the discrete case) or probability densities (in the continuous case). The requirement is that the functions $f(u)$ be strictly convex for $u = 1$. There exists a sole continuous extension by $f(0) > -\infty$ and $f(u)$ is convex in $\langle 0,\infty)$. Without breaking generality of the definition we can assume that $f(1) = 0$.

For each function $\underline{f}$ with above mentioned properties we define

$$D_f(q||p) = \sum_i p(x_i)\ f\left[\frac{q(x_i)}{p(x_i)}\right]$$

or (A.4.1)

$$D_f(q||p) = \int_{-\infty}^{\infty} p(x)\ f\left[\frac{q(x)}{p(x)}\right] dx$$

where we set $0.f(\frac{0}{0}) = 0$, $0.f(\frac{q}{0}) = q \lim_{u\to\infty} \frac{f(u)}{u}$ for $q > 0$

(the second condition follows from the requirement of continuity).

It holds that $0 \leqq D_f(q||p)$, in which the sign of equality is valid just then when $P \equiv Q$. Thus the models (X,P) and (X,Q) are similar each to the other if their f-divergence is close to zero. The maximum similarity is virtually the coincidence of the probabilities P and Q on the pertinent σ-algebra of the subsets of the sample space X. It is natural to take for maximum divergent those models in which the probabilities P and Q are orthogonal. It means that there exist disjoint subsets E, F of the sample space such that $P(E) = 1$, $Q(F) = 1$. Thus the models show maximum dissimilarity when those events have positive probability in model (X,P), the probability of which is zero in model (X,Q) and inversely.

The divergence measure $I(q||p)$, which we have introduced in Chapter 2, is one of specific f-divergences that we obtain by laying $f(u) = u\ \mathrm{ld}\ u$ in (A.4.1). Shannon and Kullback have greatest merit in recognizing it and in making use of it. Obviously $f(1) = 0$ and the function is strictly convex everywhere in its domain as $f''(u) = (\mathrm{ld}\ e)/u > 0$. Rényi was first who started to investigate also other divergences besides the I-divergence.

Now let us consider that $Q = P_{\xi\eta}$ is the joint probability distribution of a pair of random variables (ξ,η) with a sample space $X \times Y$ while P is the product of marginal probability distributions; therefore $P = P_\xi \times P_\eta$. Thus all initial mathematical objects are uniquely determined by the pair of random variables (ξ,η). We can expect the divergence between the probability $P_{\xi\eta}$ and the probability $P_\xi \times P_\eta$, which characterizes the independence of ξ and η, to determine the degree of dependence of ξ and η. We will call the quantity

$$T_f(\xi,\eta) = D_f(p_{\xi\eta} || p_\xi \times p_\eta)$$

the f-information about ξ contained in the random variable η. This terminology conforms with intuition: the greater is the probabilistic tie between ξ and η, the more information η has to bear about ξ. Because of $T_f(\xi,\eta) = T_f(\eta,\xi)$, the f-information about η contained in ξ equals that about ξ contained in η. Obviously $T_f(\xi,\eta)$ assumes non-negative values and $T_f(\xi,\eta) = 0$ only then when ξ and η are independent.

The most important of several information measures of this kind is Shannon's measure for $f(u) = u \operatorname{ld} u$, which is our above-introduced transinformation $T(X,Y)$. It is interesting to confront f-information with properties of the correlation coefficient $\rho(\xi,\eta)$, which is so far the most wide-spread measure of statistical dependence. Whereas $\rho(\xi,\eta)$ can be equal to zero even if ξ and η are dependent, the f-information has that desirable property that $T_f(\xi,\eta) = 0$ only then when ξ and η are independent. However, the behaviour of f-information in the other extreme case, namely when there exists the "highest possible" dependence between ξ and η, is more complicated and the reader is referred to literature on information theory.

It follows from what we told about $T_f(\xi,\eta)$ as a measure of statistical dependence and from mentioned properties that we can understand $T_f(\xi,\eta)$, for any particular function $\underline{f}$, as a measure of the amount of information that we obtain about the value of ξ if we observe a realization of the random variable η under assumption that both values were obtained in a random experiment controlled by probabilities $P_{\xi\eta}$. The zero information belongs to an independent experiment where the value $\eta = y$ conveys nothing about the value $\xi = x$. The greater is $T_f(\xi,\eta)$ the easier is the estimate of the value $\xi = x$ (which may not be observable directly) on the basis of the value $\eta = y$. Here we speak about the values $\underline{x}$ and $\underline{y}$ generally so that f-information is a measure that enables us to classify observations, measurements, and messages from the informative point of view but it does not tell us how informative are individual possible outcomes.

In conformity with the terminology introduced for Shannon's information we can interpret the quantity

$$T_f(x,y) = \frac{p_\xi(x)\ p_\eta(y)}{p_{\xi\eta}(x,y)}\ f\left[\frac{p_{\xi\eta}(x,y)}{p_\xi(x)\ p_\eta(y)}\right] \qquad (A.4.2)$$

as the f-information about the value $\xi = x$ contained in the value $\eta = y$. This quantity does not need be always non-negative, yet it is zero only then when the events $\xi = x$ and $\eta = y$ are independent (i.e., when

$$p_{\xi\eta}(x,y) = p_\xi(x)\ p_\eta(y)\ ).$$

$T_f(\xi,\eta)$ is then the average of these items of f-information, i.e.,

$$T_f(\xi,\eta) = \sum_i \sum_j T_f(x_i,y_i)\; p_{\xi\eta}(x_i,y_i)$$

or (A.4.3)

$$T_f(\xi,\eta) = \int_{-\infty}^{\infty}\int_{-\infty}^{\infty} T_f(x,y)\; p_{\xi\eta}(x,y)\; dx\; dy$$

Furthermore $T_f(\xi,\xi)$ yields the amount of information about ξ which we obtain if we directly observe ξ instead of another random variable loosely statistically connected with it. Thus it provides information from a direct observation of the random variable ξ. This information can be in the same time a measure of uncertainty of the random variable ξ (of uncertainty of the system described by probabilities P_ξ) so far as we understand uncertainty as difficulty in forecasting a realization of ξ on the basis of the knowledge of the probability law P_ξ. The difficulty of this forecast should be namely proportional to the information obtained by observing the random variable ξ. We call these uncertainties <u>entropies</u> with respect to historical usage. In order to be allowed to call $T_f(\xi,\xi)$ also entropy we have to load the function <u>f</u> with some restrictions to fulfil requirements laid upon entropies (finiteness, maximum value when ξ is uniformly distributed).

For y = x the function in (A.4.2) becomes

$$T_f(x) = p_\xi(x)\; f\left[\frac{1}{p_\xi(x)}\right]$$

so that we can write the entropy (using the denotation $H_f(\xi)$) for a discrete random variable

$$H_f(\xi) = T_f(\xi,\xi) = \sum_i p_\xi^2(x_i)\; f\left[\frac{1}{p_\xi(x_i)}\right] \qquad \text{(A.4.4.)}$$

where $0^2 f(\frac{1}{0}) = 0$. For continuous random variables the interpretation of f-entropy is more complicated since the quantity

$$H_f(\xi) = \int_{-\infty}^{\infty} p_\xi^2(x) \; f\left[\frac{1}{p_\xi(x)}\right] dx$$

analogous to (A.4.4) can be a measure of uncertainty for some functions, yet it does not need to hold any longer that $H_f(\xi) = T_f(\xi, \xi)$.

It is possible to approach very generally the definition of entropy as an information from a direct observation. When we observe a random event having a probability $p \in \langle 0, 1 \rangle$, it is natural to express the information contained in this observation as a non-increasing function $g(p)$ of the probability p because a very likely event bears little information while a little probable event brings much information. From this point of view the number

$$\overline{H}(\xi) = \sum_i g(p_\xi(x_i)) \; p_\xi(x_i) \qquad (A.4.5)$$

provides the average information which we gain by observing the value x of the random variable ξ. Since this information is obtained by integrating information items pertinent to individual values of ξ, we can call it, as far as we understand it as a measure of uncertainty, an integral entropy. It can be verified that f-entropy expressed as $T_f(\xi, \xi)$ is involved in this concept. For $f(u) = u$ ld u we get $g(p) = -$ ld p so that $H(\xi) = H(P(X))$ as defined in (A.1.1) for discrete random variables (an extension to continuous random variables is analogous) and is identical with T(X,X) emerging from the definition in (A.3.1).

In a similar way as above in (A.4.4) we can also define conditional f-entropy of a discrete random variable ξ conditioned by a value y_i of another discrete random variable η

$$H_f(\xi|\eta=y_i) = \sum_i p^2_{\xi|\eta}(x_i)\ f\left[\frac{1}{p_{\xi|\eta}(x_i)}\right] \qquad (A.4.6)$$

where

$$p_{\xi|\eta}(x) = \frac{p_{\xi\eta}(x,y)}{p_\eta(y)}, \qquad p_\eta(y) > 0$$

is the conditional probability function of the random variable ξ given $\eta = y$. We will be interested in <u>the average conditional f-entropy</u>

$$H_f(\xi|\eta) = \sum_j H_f(\xi|\eta=y_j)\ p_\eta(y_j) \qquad (A.4.7)$$

It can be proved that, if the function $-f(u)/u$ is convex, $H_f(\xi) \geqq H_f(\xi|\eta)$ while for individual values $\eta = y_i$ such an inequality does not need to be valid. The difference $H_f(\xi) - H_f(\xi|\eta)$ between the unconditional and the conditional entropies of a random variable ξ can be as well considered as a reasonable measure of information about ξ contained in η. Of course, a question is immediately evident: for what function <u>f</u> this difference equals $T_f(\xi,\eta)$? It can be shown that for $f(u) = u\ \mathrm{ld}\ u$ (Shannon's transinformation of discrete random variables) these two measures are equivalent (cf. Appendices

A.2 and A.3).

The concept of f-entropy can be extended to a vector $(\xi_1, \xi_2, \ldots, \xi_n)$ of random variables if we deal with multidimensional sample spaces, and the same can be done with f-information.

Index